PSA Schedule of Rates – A Training Manual

FIRST EDITION 2009

London: The Stationery Office Ltd

Published by TSO (The Stationery Office) and available from:
Online
www.tsoshop.co.uk

Mail, Telephone, Fax & E-mail
TSO
PO Box 29, Norwich, NR3 1GN
Telephone orders/General enquiries: 0870 600 5522
Fax orders: 0870 600 5533
E-mail: customer.services@tso.co.uk
Textphone 0870 240 3701

TSO@Blackwell and other Accredited Agents

Customers can also order publications from:
TSO Ireland
16 Arthur Street, Belfast BT1 4GD
Tel 028 9023 8451 Fax 028 9023 5401

First edition published 2009

ISBN 9780117068322

Printed in the United Kingdom by The Stationery Office
233569 C2.5 12/09 955 19585

Carillion Technical Publications

Carillion has many years of experience in the production of technical publications for use by construction professionals, in both the private and public sectors and is the technical author of the PSA Schedules of Rates. This experience derives from its origins within the Property Services Agency where a comprehensive portfolio of publications was developed to meet the needs of the Government Estate.

This book was written as a result of many requests from clients, maintenance managers and contractors for a simple and practical training manual for users of the PSA Schedules of Rates in a Measured Term Contract who are unfamiliar with the Schedules. It has been written to act as a companion to the book by the same author titled 'A Guide to Measured Term Contracts' (ISBN 0-11-702554-2).

Whilst the 'Guide' deals with the principles of Measured Term Contracts and provides details of how to set them up, this 'Training Manual' focuses on the interpretation of the PSA Schedules of Rates, provides advice on the measurement rules and uses a wide variety of worked examples to provide valuable guidance on how to measure items based on the Schedules.

Carillion is an innovative and quality driven business providing a wide range of consultancy services that include:

- bespoke schedules of rates and specifications;
- measured term contract advice and training;
- cost consultancy and professional quantity surveying;
- financial, technical, procedural and value for money audits;
- building surveying;
- insurance reinstatement cost assessments;
- building control and fire consultancy;
- energy and environmental management.

To find out more about Carillion's consultancy services, please contact:

Carillion Schedules of Rates Group
TPS
17th Floor, Centre Tower,
Whitgift Centre,
Croydon, Surrey,
CR9 0AU

Telephone: 0870 128 5220
Facsimile: 0870 128 5204
Email: scheduleofrates.psa@carillionplc.com

www.carillionplc.com

Computerised Estimating and Term Contract Administration Systems

The following firms have been granted a licence to reproduce the PSA Schedules of Rates in a computerised form:

Barcellos Limited
Sandbach House
8 Salisbury Road
Leicester LE1 7QR

Telephone: 0116 233 5559
Facsimile: 0116 233 5560
www.barcellos.co.uk

Qudos Computer Software Limited
Ashmead House
3 The Common, Siddington
Cirencester, Glos. GL7 6EY

Telephone: 01285 656812
Facsimile: 01285 655270
www.qudossoftware.com

MTC Updating Percentages Online

These monthly updating percentages may be used as a contractual basis for the reimbursement of increased costs for measured term contracts let on all PSA Schedules of Rates or other forms of maintenance contracts as well as assisting in updating estimates.

Available from:

Building Cost Information Service
12 Great George Street
Parliament Square
London
SW1P 3AD

Telephone: +44(0)20 7695 1500
Email: contact@bcis.co.uk

Order online at www.bcis.co.uk/online (subscription service).

Contents

1 Introduction

The Management of Maintenance Contracts has in the past been treated as a matter of little importance mainly because the value of individual orders is very small. However the overall volume of maintenance work let by many organisations in one year often outweighs their expenditure on new work. A more vigorous approach to the management and cost control of maintenance work is therefore necessary.

Measured Term Contracting is one of the methods of obtaining value for money in maintenance open to the manager. It has distinct advantages over "cost plus" type contracts in that the price paid to the Contractor is defined in a schedule of rates. The Contractor carries the responsibilities of his own inefficiencies, there is less need for site supervision and, if administered properly, less opportunity for disagreement of final accounts.

The use of Measured Term Contracts (MTCs) has increased rapidly in recent years as government initiatives have led local authorities and others to introduce competitive tendering for maintenance work.

The Schedule of Rates is the key component in a Measured Term Contract as it is used to determine the successful tenderer and forms the contractual basis on which payment is made. This Training Manual is designed to give measurement and pricing guidance for users of the PSA Schedule of Rates who are unfamiliar with the Schedules, including descriptions of the contents and format, advice on measurement rules, worked examples and frequently asked questions.

2 History

The Property Services Agency (PSA) and its forefathers had managed the construction and maintenance of Public buildings for over 600 years, but it was not until 1865 that the first known Schedule of Rates was published. The old War Department Schedule, which included 10 trades and over 7,000 prices, was used in many corners of the world for maintenance of the Defence estate. Building and maintenance work could be valued at pre-determined prices leading to the gradual emergence of the Term Contract.

However the most sweeping move towards Measured Term Contracts in the Public sector came 100 years later in 1965, when the newly formed Ministry of Public Building and Works took over responsibility for maintenance of the whole of the Defence works in addition to the Civil Estate. The Ministry recognised the advantage to be gained from a series of 'call-off' contracts or standing arrangements whereby a heavy programme of maintenance and small works could be ordered at pre-set prices without the need for raising a separate contract on each occasion.

Since then the Measured Term Contract has become the cornerstone of the maintenance operation and although it is not without problems, if used appropriately it will serve the user well.

3 What is a Measured Term Contract?

The Measured Term Contract is an arrangement whereby a Contractor undertakes to carry out a series of Works Orders, over a period of years, within a defined geographical area and where the work is subsequently measured and valued at rates contained in a pre-priced Schedule of Rates.

In order to obtain competition, tenderers submit a contract percentage on or off the Schedule of Rates to reflect their assessment of all additional factors (such as overheads, market conditions, etc). Sometimes, separate percentages are requested for different value bands or categories of work.

The Contract percentage is subsequently applied to all work priced against the Schedule of Rates. Where one of the PSA Schedules of Rates is used, a second adjustment is made - to update the pricing level of the Schedule of Rates to the current month, using published "MTC Updating Percentages Online" available from the Building Cost Information Service (BCIS). These monthly updating percentages may be used as a contractual basis for the reimbursement of increased costs for measured term contracts let on all PSA Schedules of Rates as well as assisting in updating estimates.

A Measured Term Contract is appropriate in the following circumstances:

1. Where the Employer has an ongoing need for maintenance/minor new work. The contract period is generally a one to three-year term, although longer periods of up to five years have been known for specialist contracts, e.g. Landscape Management.

2. Where there is sufficient workload to offer continuity and economy. The workload should be large enough to attract Contractors, and to offer them continuity of work, which in turn should provide the Employer with a saving in cost and documentation.

3. Where a prompt response is required. The Measured Term Contract provides the opportunity for a quick response but needs good communication and management to achieve a consistent response over a period of three years or more.

The user also needs to consider the availability of suitable Contractors and any particular requirements of the Employer.

The Measured Term Contract is essentially a package. It will not necessarily provide the lowest price for each individual Works Order but it will usually produce the best overall solution to an Employers' maintenance needs.

The price paid is a classic case of 'swings and roundabouts'. If a Contractor has to replace all the rainwater goods on a housing estate the Schedule of Rates will reimburse him handsomely - on the other hand if he is required to travel 50 miles to replace a pane of glass he will not be so fortunate. It is important therefore, to order all work covered by the Schedule of Rates from the Contractor - not just the low value and awkward work. The Contractor will have tendered on the basis of all relevant work with a spread of value.

Types of Measured Term Contract

The various Measured Term Contracts used nationally relate to the various Schedules of Rates available on the market. In the PSA, the Building and Civil Engineering Schedule of Rates was the largest containing about 20,000 items in tabular form and accounting for over 40% of the Measured Term Contract workload. There are other Schedules of Rates widely used to cover the vast majority of maintenance and minor new work the user is likely to experience.

TABLE 3.1 MEASURED TERM CONTRACT TYPES USED BY PSA

		Approx. Annual Spend (£)	%
1.	Building Works	121	43
2.	Decoration work	23	8
3.	Grounds maintenance	33	11
4.	Roadworks & paving	24	8
5.	Electrical services	53	19
6.	Mechanical services	28	10
7.	Maritime services	2	1
8.	Railway works	1	-
		£285M	100%

Constituent elements of a Measured Term Contract

The main constituent elements of a Measured Term Contract are:

1. The Schedule of Rates
2. The Contract Conditions
3. The Particular Clauses

The Schedule of Rates

A Schedule of Rates consists of a list of unit items of work priced at a rate per unit (e.g. per m^2, kg, etc.). The items of work are fully described and generally include specification clauses. Usually, Schedules of Rates do not include quantities. Dependant on the type of Schedule of Rates used, the list of items will either be pre-priced by the Schedule writer or priced by Contractors during the tendering process. The Schedule of Rates thus formed is used in conjunction with the measurement of work to calculate payment. Schedules of Rates are explored in greater detail in Chapter 5.

The Contract Conditions

The Contract Conditions are the "small print" and define the rights and responsibilities of the two parties to the contract and such details as the responsibility for measurement and periods of payment. Some employers write their own contract conditions but it is usual to use a Standard Form produced by institutions or organisations with a vested interest in contracts that will produce the minimum of disputes. Standard Forms have the advantage of familiarity, as many Contractors, consultants and Employers will have used them before. Users may also have developed established practices and procedures in line with Standard Forms. In addition, it is also obviously expensive to have a contract written from first principles using legally qualified staff.

The Particular Clauses

Particular clauses usually relate to scope of the work, overtime, sub-contracting, and any Employer, or other, requirements and particular site circumstances such as:

- security arrangements
- service installations
- 24 hour/365 days a year emergency call-out service
- 24 hour/365 days a year on-site attendance service
- Specialist sub-contractors.
- Minimum and maximum response times

4 Advantages and Disadvantages of Measured Term Contracts

Advantages

1. Flexibility/Accountability

The cost of work can be evaluated before a Works Order is raised thereby providing a reliable and accurate estimate. Expenditure and commitment is easily monitored providing close financial control. There is no more flexible method of procuring small works, while retaining accountability, whether it is to the Public Accounts Committee, the taxpayer or the shareholders.

2. Minimal documentation

A multitude of Works Orders may be raised under the one contract with very little in the way of documentation. Tender action is only required once each term.

3. Immediate response

Without the requirement for contract action, the Contractors response should be immediate once a Works Order is issued.

4. Saving in time and resource cost

Resulting from pre-contract documentation and post-contract response time.

5. Contractor 'on call'

Apart from being in a position to respond quickly to normal demands, the Contractor should be able to deal with any emergency work that is needed.

6. Familiarity/Continuity

The Employer benefits from the Contractor's increasing familiarity with the buildings while the Contractor benefits from the assurance of a reasonable continuity of work. Once given a selection of Works Orders the Contractor can programme his work to make the most efficient use of his resources.

There is an excellent opportunity for both parties to create a good long-term working relationship.

7. Value-for-money

Measured Term Contracts provide extremely good value-for money. The PSA carried out an exercise of re-pricing over 100 lump sum contracts (up to £25,000 in value) on the basis of available Measured Term Contracts. Results showed that overall, taking account of 'swings and roundabouts', there was little or no difference in pricing levels. There may be a potential saving by use of lump sum contracts for the higher value Works Orders but the lower value work - which forms the bulk of Measured Term Contract spend - showed a clear advantage to the term contract.

8. Termination

The termination clause is fair to both parties offering a remedy for poor performance or a change in circumstances.

9. Simplicity

Tender documentation, adjustment for inflation and the ordering of work are all simple operations.

Disadvantages

1. Volume of Measurement

Frequently there is a comparatively large amount of measurement for the small value Works Order - hence the move towards self-measurement, which may itself require closer control.

2. Delayed accounts

There is sometimes insufficient incentive for the Contractor and Quantity Surveyor to be up-to-date with measurement and valuation, particularly where the Contractor has received advance payments. This situation has been recognised and the latest contract conditions address the problem. Nevertheless regular monitoring is recommended.

3. Familiarity/convenience

Once a term contractor is established it becomes very convenient to use him for work outside his remit and this can overtake commercially sound reasons to adopt other procurement methods.

4. Availability of suitable contractors

Generally, term contracts require Contractors with considerable all round expertise and this can limit the availability of suitable Contractors.

5. Audit requirement

The volume of separate Works Orders and high value of Measured Term Contract spend does create an auditing problem.

The Employer may need to set up an auditing regime to determine if they are receiving value for money. This is more likely to be necessary if;

- A large proportion of the works are Contractor self measure on a detailed Schedule of Rates
- The employer is concerned about fraud between the Contractor and Employer's ordering staff.

5 Schedules of Rates

Introduction

The Schedule of Rates is the key component in a Measured Term Contract as it is used to determine the successful tenderer and forms the contractual basis on which payment is made. Schedules of Rates are also used in conjunction with lump sum contracts to provide a basis for valuing variations to the contract, but this is a separate topic in its own right. They can also provide a benchmark for comparing regional variations and monitoring tendering trends

This Chapter describes the role of the Schedule of Rates within a Measured Term Contract, the contents of a Schedule, the types of Schedule available and the choice and preparation of Schedules. It is worthwhile at this stage to make the distinction between "Schedules of Rates" and "Price Books".

There are many Price Books available but these are used, primarily, for estimating building works. They could, theoretically, be used as Schedules of Rates but published Schedules have the advantage of the inclusion of measurement rules and specification notes, the availability of compatible computerised administration systems and historical links with published Standard Forms of Contract.

It is important to the success of a Measured Term Contract that an appropriate Schedule of Rates is provided. Time spent in choosing or preparing a Schedule of Rates to suit particular circumstances will be more than repaid in savings during the operation of the Contract.

Role of the Schedule of Rates

The Schedule of Rates fulfils the same two primary roles as a Bill of Quantities fulfils for a lump sum contract, i.e. it provides the basis for tender assessment and the basis for payment. It also fulfils some secondary roles in facilitating budget planning, cost control and ordering.

Primary Functions

1. Defining specification standards

A Schedule of Rates can be prepared to include materials and workmanship specification clauses. These are either incorporated in item descriptions or separate preamble clauses. Although the Project Manager/Contract Administrator needs to prepare a thorough description of the work required for an order, the specification clauses serve to prevent the scope, standard and cost of work escalating. The comprehensiveness and stand-alone nature of the specification clauses are factors that make pre-priced standard Schedules of Rates ideal for use as tendering and contract documents.

2. Tender assessment

The Schedule of Rates allows a comparison between Contractors based on price. This comparison can be very straightforward if the tender requires a single tendered percentage for all work. It is slightly more complicated if the tender requires separate percentages for different trades, locations, value of Works Orders, etc. Each percentage will need to be weighted according to frequency of occurrence. However, every rate will be different if a blank Schedule has been issued for pricing by Contractors and, consequently, assessment would be extremely time-consuming.

3. Basis of Payment

The priced Schedule of Rates and its attendant percentage additions or deductions, allow for work to be carried over the period of the contract at an agreed price, even though the exact composition of the work and its timing are not known at the time of tendering. This use of the Schedule is the basis of Measured Term Contracting.

Subsidiary Functions

1. Budget Planning

The priced Schedule of Rates enables managers to estimate with a degree of accuracy the cost of their programme and to compare alternative programmes at an early stage.

2. Cost control of the budget

The ability to value the Works Orders as they are issued at agreed rates allows managers to know the amount of work committed and therefore to control the flow of work to meet budget requirements.

3. Ordering of Work

Certain types of Schedules of Rates can form the basis of a Works ordering system. For example, a Composite Schedule of Rates may contain descriptions of complete tasks such as “Forming a new opening in a wall, providing a new window, glazing, plastering the reveals and painting”. Such a description could be referred to on the Works Order form by its “item number” in the Schedule. This simplifies administration as the invoice will exactly match the Order and reduce the need for measurement and checking.

Contents of a Schedule of Rates

The exact contents of the Schedule of Rates will vary according to requirements of the individual contract, but Schedules would normally contain preliminaries, preambles and priced items of work.

Preliminaries

The preliminaries should define the scope of the contract, detail the conditions of contract and clarify any matter that the Employer wishes to bring to the Contractor's attention. Where a standard published Schedule of Rates is used, the preliminaries may take the form of a separate document.

Where a standard form of contract is used the preliminaries only need to contain the clause headings and additional special conditions. If the Employers own procurement contract is being used the preliminaries may well contain the detailed contract conditions.

Typical matters that may need to be covered include:

1. Scope of the contract - including name of the employer and supervising officer.

2. Provision for determination of the contract by either party.

3. Method of ordering work and the time for completion.

4. Liquidated damages for delayed completion.

5. Compliance with statutory regulations.

6. Plant, tools and vehicles; provision and maintenance of scaffolding.

7. Responsibilities for the provision of materials including materials supplied by the Employer.

8. Consent required for subletting of the contract.

9. Access to the site or property by the Contractor - including provision to enable the Contractor to carry out the work.

10. Valuation of the work - in accordance with the Schedule of Rates and the percentage adjustment.

11. Agreement of rates where prices in the Schedule of Rates may not apply - including provision for dayworks.

12. Interim payments on monthly valuation.

13. Certification for payment by the supervising officer of the Contractor's account within a specified period.

14. Defects liability - provision for the Contractor to make good any defects appearing within a specified time after completion of the work.

15. Materials, works and workmanship to conform to description.

16. Removal of rubbish.

17. Insurance of the works against fire, etc and insurance of existing structures.

18. Insurance against damage to property and injury to persons - third party insurance.

19. Assignment of the contract.

20. Value Added Tax.

21. Methods of settling disputes.

22. Contract definitions - the Contract, Employer, Contractor, Contract Area, Contract period, Works Order or Orders, Site, Schedule of Rates, Contract Administrator/Project Manager, Percentage Adjustments.

Most of these items are covered by Standard Forms.

Preambles

Sufficient specifications preambles should be included to define the materials and workmanship required. They should not be over elaborate and should reflect the nature of maintenance work. There is sometimes a tendency to pad out the preambles with masses of specifications for new work when the Schedule of Rates only has items for repairs. For example there is no point reproducing pages of new work plastering specifications if all that is included in the Schedule is patch repairs to plaster and making good.

Priced Rates for Items of Work

The Schedule of Rates should contain the descriptions of the work and the price to be paid, subject to tendered percentages. Some approximation of the annual frequency of each item could be included. This information is rarely available but it would undoubtedly be useful to Contractors if it were.

It is normal to issue a Schedule of Rates with all inclusive prices but in the past Measured Term Contracts have been run using Schedules with labour only rates leaving materials costs to be reimbursed at cost.

There have been some instances of blank Schedules of Rates being issued at tender stage for Contractors to price. This requires a lot of work by the tenderers and can lead to some problems at the tender assessment stage, as every rate for every tenderer will be different. This makes pricing blank Schedules fairly risky for both the Contractor and the Employer.

When published Schedules of Rates do not cover particular regularly occurring work, consideration should be given to producing a local addendum.

Attributes of a good Schedule of Rates

A Schedule of Rates should be concise, complete, clear and consistent.

Concise:

In content - The Schedule should not contain items that will never be used. The inclusion of unnecessary items will mislead the Contractor as to the nature of the work and make the document unwieldy in use. It is pointless to include rates for new work or, for example, areas of roof tiling if the work to be ordered will only ever be the replacement of tiles in ones and twos.

In presentation - The Schedule should avoid repetition by use of headings for details which are common to a succession of jobs and by not attempting to differentiate between types of work which are not significantly different in nature or price.

Complete:

The Schedule should cover all the types of work that need to be carried out under the contract. It is impossible to avoid some dayworks on a maintenance contract but the drafting of the Schedule should endeavour to keep this to a minimum.

There obviously has to be some trade off between conciseness and completeness. However a working knowledge of the estate to be maintained should indicate the scope of the items to be included. The 'We will stick it in just in case' approach should be avoided.

If the scope of the Schedule does reflect the work that is to be carried out it will help the Contractor in understanding the nature of the work for which he is tendering and should lead to better value for money for the Employer.

Clear:

The items should be unambiguous as to specification, shape, size, etc. The descriptions should define the work included so that arguments do not arise during the contract. The unit of measurement should also be unambiguous. The use of a Standard Method of Measurement will assist definition.

Consistent:

The pricing level should be consistent throughout the Schedule. It makes it difficult for Contractors to tender a single percentage on a Schedule if the rates within each trade, or even between trades, are unevenly priced. No two surveyors or Contractors will price a selection of items the same. What may seem to be consistent pricing to one person will be considered uneven to another, but obvious inconsistencies should be avoided.

The pricing should also be a reasonable market price. In theory the Contractor is pricing independently of the Schedule but there must be some doubts raised if the rates presented require a large adjustment. Low rates requiring a big addition may well make the Contractor suspicious of the whole exercise and while Contractors will and do tender negative percentages, there is probably a psychological barrier to putting too large a negative percentage in a tender.

Types of Schedules of Rates

An ideal Schedule of Rates would be one that contains the exact description of the work to be carried out on every Works Order. Given the range of possible items and the possible combinations of items, the permutations involved are probably infinite. There are two possible solutions to this problem and it is these two solutions that define the two main types of Schedule of Rates available:

- Detailed Schedules, and;
- Composite Schedules.

The Detailed Schedule of Rates contains individual items of work for separate tasks. This allows a job to be priced by building up from individual items of work, very much like a traditional estimating price book.

The Composite Schedule of Rates contains tasks built-up from the individual items of work involved in carrying out the task.

For example, if the actual work involves replacing a wash hand basin including replacing the wall tiles in the splashback, resealing the basin and overhauling the taps, the Detailed Schedule will have separate items for replacing the basin, replacing the tiles, sealing the basin and overhauling the taps. The Composite Schedule may have a single item for replacing a wash hand basin including all making good to tiles and plaster, overhauling taps and renewing back nuts and traps where necessary.

Advantages and Disadvantages

There are advantages and disadvantages to both types of Schedule of Rates.

Detailed Schedule of Rates

Advantages

1. There is greater accuracy of payment as the value of the Order is for work actually carried out and, therefore, there is less risk to the Contractor. This may possibly produce keener pricing.

2. There is increased flexibility as the larger number of individual items of work allows for variations in the Work Order and the Schedule of Rates is more likely to be applicable in different situations.

Disadvantages

1. There is a greater requirement for measurement and valuation and this will increase administration and, possibly, overall costs.

2. There is less accuracy in estimated values of Works Orders. These will be 'guesstimates' unless pre-measurement is carried out.

3. They are not suitable for providing the basis of Works Orders without pre-inspection.

Composite Schedule of Rates

Advantages

1. There are reduced administration costs as there is less requirement for pre- or post-measurement. Descriptions of tasks on the Works Orders are more likely to be the same as those in the Schedule.

2. They are quicker and more economical in situations where Works Order values are low. Building up low value Works Orders from several individual items is not cost-effective.

3. There is greater certainty of committed money as the Works Order estimated value and final value will be similar or even identical.
4. The composite descriptions can be used for placing Work Orders. This reduces the need for an exact diagnosis of faults by the person issuing the Order as the repair item will cover a variety of possible tasks required to solve a problem. It is, therefore, not necessary to predetermine which task will be required when ordering the work.

5. The Composite Schedule will help to define the work. The Schedule will indicate the scope of work likely to be included in the contract.

Disadvantages

1. It lacks flexibility in that it has limited use if work beyond the original scope of the contract arises. Furthermore, the Schedule is not portable in that it cannot be used on a different estate or by a different Employer without significant amendment.

2. There is a pricing risk in that the pricing requires some averaging of the likely work to be carried out and the resultant rates will rely on 'swings and roundabouts'. This increases the risk to the Contractor, which will be reflected in his tender and also introduces some risk to the Employer.

Choice of Schedule of Rates

There is a trade off between increased accuracy of payment and increased cost of measurement and administration in choosing between these two types of approach.

The two main factors affecting the choice are the nature of the estate and the likely value of Works Orders. If the estate contains a fairly narrow band of building types with similar recurring maintenance problems, a Composite Schedule of Rates would seem more appropriate. If the estate to be maintained is likely to produce a wide variety of different items of work, because of the nature of the buildings or the variation in their specification or use, then a Detailed Schedule of Rates is more appropriate. Housing Schedules of Rates should be as simple as possible providing identification of complete jobs rather than items of work.

If the average value of the Works Orders is small the cost of measurement or checking becomes prohibitive. However if the average Works Order value is large, particularly if it is likely to contain more than one job, the post measurement becomes a requirement and the increased accuracy obtained from a Detailed Schedule of Rates probably outweighs the administration costs.

Table 5.1 shows the factors that would tend to make one type of Schedule of Rates more or less applicable.

Table 5.1 Choice of Schedule of Rates

Detailed Schedule	Composite Schedule
High average Works Order value	Low average Works Order value
Disparate building types	Uniform building types
Varied Specification	Consistent Specification
Varied usage of building	Consistent usage of building

It is possible to mix the two approaches to provide composite rates for regularly occurring items and detailed rates for less predetermined work but care must be taken in mixing detailed and composite rates to ensure that the composite items are clear in their intent and do not encourage the Contractor to add detailed rates for work that should be covered in the composite item. Nor should the Contractor be left to choose whether to use the composite item or the detailed measurement.

Availability of Schedules of Rates

In setting up a Measured Term Contract there are two broad options for acquiring an appropriate Schedule of Rates:

- use a published Schedule of Rates; or
- have a bespoke Schedule of Rates tailor made.

There are advantages and disadvantages with each solution.

Published Schedule of Rates

Advantages

1. Familiarity - many Contractors' consultants and Employers will be familiar with the Schedule, which could produce more confidence among tenderers and consequently keener tenders.

2. Availability - they are readily available. They can be obtained from specialist bookshops or direct from the compilers.

3. Established Practices - they may have existing procedures developed for operating term contracts or there may be computer programmes available to use with the Schedules.

4. Format - they tend to be detailed rather than composite, which is an advantage if a Detailed Schedule is required.

5. Cost - relatively cheap for a control document for fairly large contracts.

It is worth bearing in mind that a Bill of Quantities for a £1,000,000 new work contract will cost something in the order of £10,000 to £15,000. A Measured Term Contract with a similar volume of expenditure should be regarded as in need of equally valuable documentation. The cost of a published Schedule of Rates and associated administrative documentation are small compared with the benefits.

Disadvantages

1. Loose-fit - they will not be concise and will contain items that are not appropriate to a particular estate - this may make them unwieldy.

2. Non-Specific - they do not help to define the type of work to be carried out by the Contractor.

3. Format - they tend to be detailed rather than composite, which is a disadvantage if you want a Composite Schedule.

4. Compatibility - may lack compatibility with existing procedures and programmes.

Bespoke Schedules of Rates

Advantages

1. Concise - can be concise and specific to the estate and therefore easier to use.

2. Format - can be detailed or composite as appropriate.

3. Specific - help to define the work to be carried out by the Contractor.

4. Ordering of work - They can be designed to facilitate the placing of Works Orders.

5. Compatibility - They can be designed to fit with existing procedures and computer systems.

Disadvantages

1. Unfamiliarity - there will be a learning curve for both the Contractor and the Employer.

2. Cost of production - the cost will obviously vary depending on the range of work and number of rates, but something similar to the cost of preparing a Bill of Quantities will give an indication. Whatever the cost it will be negligible compared with the amount of money spent using it.

3. Availability - it will take some time to prepare.

There is, of course, an option to mix the two using the published Schedule of Rates together with a bespoke Schedule of Rates for more particular items of work.

PSA Schedules of Rates available

The Schedules are published approximately every five years. They are comprehensive, cover new work as well as maintenance and are clear and detailed in their specification. The presentation tends to be tabular and is available in an electronic form. A bespoke service is offered whereby Employers can have a Schedule of Rates prepared based on their particular requirements.

Since the privatisation of the PSA, the Schedules of Rates have been produced by Carillion Services and published by The Stationery Office. The current suite of commercially available Schedules comprise:-

Building Works - this covers all aspects of general building work. It includes approximately 21,000 work items/rates.

Mechanical Services - this provides comprehensive coverage of all aspects of mechanical services. It includes approximately 11,000 work items/rates.

Electrical Services - this is a companion volume to the Mechanical Services Schedule, and covers all aspects of electrical services. It includes approximately 10,000 work items/rates.

Decoration Works - this is applicable to complete packages of decoration and redecoration works.

Landscape Management - this covers all aspects of grounds maintenance, hard and soft landscaping and street furniture.

The rates are calculated from first principles and take account of labour constants derived from research and historic data, industry accepted labour rates, materials costs derived from manufacturer's costs/suppliers price lists (adjusted for trade discounts as appropriate) and plant costs derived from research and historic data. The rates also include an allowance for preliminaries (site overheads). The rates are all expressed at the price levels applicable to the base date of the Schedule. The base date is the reference point used for updating the rates in the Schedule.

The Building Cost Information Service (BCIS) independently produces and publishes percentages specifically for the purpose of updating the PSA Schedules of Rates.

Fluctuations

Whichever Schedule of Rates is chosen there will be a requirement to allow for increased costs where contracts are let over a period longer than a year.

The PSA Schedule of Rates use the application of "MTC Updating Percentages Online" available from the Building Cost Information Service (BCIS). They provide a simple way of bringing the Schedule of Rates up to current price levels, and are generally accepted by the Building Industry. They are produced monthly, by applying the increased cost of labour, material and plant to a statistical sample of items in the Schedule of Rates. The sample of items is re-assessed when a new Schedule is produced.

6 PSA Schedules of Rates

The following is based on the PSA Schedules of Rates for Building Work 2009, the PSA Schedules of Rates for Electrical Services 2006 and the PSA Schedules of Rates for Mechanical Services 2006, but the basic principles apply to other editions.

Contents of the Schedules

- General Directions
- Work Sections
- Appendices
- Index

Format of the Schedules

Apart from the General Directions Section, the Schedule is structured in accordance with the "Common Arrangement of Work Sections For Building Works" (Second Edition) published in 1998 by the Construction Project Information Committee as part of the initiative towards Co-ordinated Project Information

The Schedules contains Work Groups labelled from A to Z

Each Work Group is divided into Work Sections

The Work Groups for the Building Works Schedule are as follows:
A General Directions
C Demolition, Alteration and Renovation
D Groundwork
E In-situ Concrete
F Masonry
G Structural and Carcassing Metal and Timber
H Cladding and covering
J Waterproofing
K Linings, Sheathing and Dry Partitioning
L Windows, Doors and Stairs
M Surface Finishes
N Furniture and Equipment
P Building Fabric Sundries
Q Paving, Planting and Fencing
R Disposal Systems
S Piped Supply Systems
Z General Items

The Work Groups for the Mechanical and Electrical Schedules are as follows:
A General Directions
S Piped Supply Systems
T Mechanical Heating and Cooling Systems
U Ventilation and Air Conditioning Systems
V Electrical Supply Systems
W Communications, Security and Control Systems
Y Services

Items are referred to by their "Item Reference", e.g. 9.5 mm Gypsum plasterboard lining to wall: 2400 mm high (page 200 of the Building Works Schedule) is "K10.001/1".

Items can be found by finding the sub-section in the Contents or looking up the description in the Index.

General Directions

- Preliminaries
- Sub-Contracts
- Method of Measurement
- Basis of the Rates

The Rates are all inclusive and as such, include for Preliminaries.

Significant preliminary items

When assessing any adjustment to be tendered to the level of the Rates, consideration must be given to any unique preliminary items which could be encountered

The following list gives examples only and does not limit the actual requirements encountered:

- Establishment charges, overheads and profit
- Nature and location of the Works
- Liabilities and insurances
- Access to the Works
- Existing services
- Plant, tools, vehicles and transport
- Site organisation, security, health and safety
- Water, lighting and power
- Temporary works: e.g. hardstandings, accommodation, storage, telephones, fencing, footways and gantries
- Statutory obligations

Sub-Contractors and Suppliers

The Contractor, unless the Conditions of Contract state otherwise, is reimbursed the net agreed amount of the account (after the deduction of all discounts obtainable for cash, insofar as they exceed 2½ per cent, and of all trade discounts, rebates and allowances) with the addition of 5 per cent to cover profit and all other liabilities reimbursement is not subject to any further adjustment.

Ensure that all discounts and allowances (save 2½% cash discount) are deducted. Is the Contractor entitled to annual rebates from suppliers not shown on invoices? Is a photo-copy invoice a true copy of the original? Beware of using prices from invoices which contain only small quantities

Method of Measurement

The Schedules stand alone. Where it is not self-evident from the item descriptions, the rules of measurement are stated. Every effort has been made to adhere to the Standard Method of Measurement of Building Works: Seventh Edition (SMM7), but by the very nature of the Schedule's uses some expansion is necessary.

Work is measured nett as fixed in position unless otherwise stated. No allowance for waste should be made when measuring as an average allowance for waste is included in the Rates.

Take dimensions used in calculating quantities to the nearest 10 mm, e. g.

1.683 m would be recorded as 1.68
3.457 m would be recorded as 3.46

Bill all quantities to the nearest two decimal places of the billing unit for each particular item.

Size ratings expressed as "...... to ...." are read as "exceeding .... but not exceeding ...."

The Rates

The Rates in the Schedule reflect the costs of resources as at the Base Date stated under "Rates" in the General Directions.

Adjustments to the Rates can be made using the MTC Updating Percentages Online, available from the Building Cost Information Service. These monthly updating percentages may be used as a contractual basis for the reimbursement of increased costs for measured term contracts let on all PSA Schedules of Rates or other forms of maintenance contracts as well as assisting in updating estimates.

Where two or more multiplying factors are to be applied to a rate, the factors are first multiplied together and not added.

The Building Works Schedule rates

What the Rates exclude:

- Work normally executed by steeplejacks.
- Work carried out in compressed air
- Work carried out in or under water
- Preparing drawings.

What the Rates include:

All Rates:

- Preliminaries, labour, waste and lost time.
- Supplying and delivering materials, unloading and getting into store or other approved position.
- Storage of materials.
- Assembling, fitting and fixing materials in position.
- Carrying out work in any circumstances. Work carried out in disadvantageous circumstances (e.g. work in occupied buildings, alterations or extensions to existing installations, etc.) will be offset by work carried out in advantageous circumstances.
- Work at any height unless otherwise stated.
- Scaffolding or working platforms not exceeding 4.50 m above the base.
- Executing work in sections as necessary.
- Protecting old and new works from damage.
- Keeping the site clean and tidy.
- Leaving the site clean and tidy upon completion.

Taking out or taking up existing:

- Removing from the building
- Removing off site
- Paying any disposal charges
- Removing to stack or storage place on site

Fixing only or laying only:

- New items or items previously set aside for re-use
- Taking delivery and storing new items
- Obtaining from stack or storage any items
- Supplying fixing or jointing materials

Work in repairs:

- Removing any existing work
- New work to match existing

- Preparatory work and making good
- Jointing new to existing
- Work of any width

The Mechanical and Electrical Services rates

There are five rate categories:

- Rate A: Supplying materials.
- Rate B: Installing.
- Rate C: Supplying and installing.
- Rate D: Dismantling and removing for scrap.
- Rate E: Dismantling and removing for re-use.

Rates A, B and C are as published.

Rates D and E are derived by applying the appropriate multiplying factor to the principal items being dismantled as stated in the particular Section preambles - ancillary and related items not dismantled from such principal items are deemed to be included.

What the Rates exclude:

- Electrical work (in the case of the Mechanical Schedule)
- Mechanical work (in the case of the Electrical Schedule)

What the Rates include:

All Rates:

- Preliminaries, labour, waste and lost time.
- Supplying and delivering materials, unloading and getting into store or other approved position.
- Storage of materials.
- Assembling, fitting and fixing materials in position.
- Carrying out work in any circumstances. Work carried out in disadvantageous circumstances (e.g. work in occupied buildings, alterations or extensions to existing installations, etc.) will be offset by work carried out in advantageous circumstances.
- Work at any height not exceeding 4m from a firm base
    - 4 to 7 metres - multiply by 1.25
    - 7 to 10 metres - multiply by 1.45
    - exceeding 10 metres - multiply by 1.75

Rate A:

- Supplying, delivering, unloading, getting into store or other approved position, items including materials (except materials included under Rate B, e.g. jointing and fixing materials), component parts, equipment, associated installation and servicing tools, templates and handbooks.
- Works testing and certifying.

- Providing certificates of compliance.

Rate B:

- Preparing items for installation, disposing of packaging or the like.
- Any necessary storage.
- Moving items to position of installation and returning defective items to manufacturer and surplus items to store.
- Gaining access to installation position including lifting and replacing removable duct, conduit or trunking components, duct covers and the like.
- Installing items in any location or position and fixing to any background by whatever means are necessary including any assembling and dismantling required for the purposes of installation.
- Assembling, cutting, jointing and connecting.
- Supplying fixing materials, including fixing clips or the like for which rates are not shown separately.
- Supplying heat and jointing and connecting materials.
- Temporarily fixing in position and taking down as many times as necessary prior to permanent installation.
- Notching or boring timbers or wall board deemed necessary for fixing.
- Installing new or old items previously set aside for re-use.
- Preparing old work to receive new.
- Marking out and/or indicating any other building work required.
- Isolating and re-connecting whole or parts of existing systems as appropriate including ensuring that existing installations (where affected but not isolated) continue to function during the execution of the works.
- Making good and leaving suitable for permanent reinstatement and/or decoration.
- Carefully examining the installation to verify that all materials comply with the ordered requirements, inspection and testing of work installed and for adjusting and setting to work. (For definitions of terms and items deemed to be included, see section Y81 Testing and Commissioning).
- Removing and disposing of rubbish and debris and leaving clean on completion.

Rate C:

- Everything included under both Rate A and Rate B.

Rates D and E:

- Carefully taking out including all cutting, disconnecting, unscrewing, unbolting and dismantling.
- Providing all necessary heat.
- Dismantling as directed in excess of that necessary for scrap (Rate E).
- Carefully taking out all fixings from the item removed and from the background.
- Isolating and re-connecting whole or parts of existing systems as appropriate including ensuring that existing installations (where affected but not isolated) continue to function during the execution of the works.

- Getting out items to be scrapped (Rate D) and removing and disposing of in a designated or authorised place.
- Removing to store items to be re-used (Rate E).
- Cleaning and preparing items for re-use (Rate E).
- Gaining access to existing installations including lifting and replacing removable duct, conduit or trunking components, duct covers and the like.
- Making good and leaving suitable for permanent reinstatement and/or decoration.
- Removing and disposing of rubbish and debris and leaving clean on completion.

Measurement Rules

Measurement Rules set out:

- When work shall be measured
- Method by which quantities are computed
- What is deemed to be included

Every effort has been made to adhere to the Standard Method of Measurement of Building Works: Seventh Edition (SMM7), but by the very nature of the Schedule's uses some expansion is necessary.

Consult:

- "Generally" Section at the beginning of each Work Section.
- "Rates for the following include" at the beginning of each Sub-Section.

7 Setting down dimensions

It is recommended that dimensions are set down as shown in the Examples.

Column 1 is the "timesing" column. If it is found that, after dimensions have been recorded, there are more than one of the same item, then the item is "timesed" (see "Earthwork Support" in Example BW23. The dimensions are the "mean girth" of the trench and the depth of the trench, but the dimensions are "timesed" by two for the two sides of the trench).

Column 2 is the "dimension" column. Dimensions are set down immediately under each other and separated from other items by a line.

- Cubic measurements will have three dimensions (length, width, depth)
- Square measurements will have two dimensions (length, width)
- Lineal dimensions will have one dimension (length)
- Enumerated items are shown as a number.

Column 3 is the "squaring" column. This will show the result of each calculation (volume, area, etc.).

Column 4 is the "description" column. A description of the work to which the dimensions apply. It is traditional to use abbreviations to save time.

Column 5 is the "item reference" column. It is important to get into the habit of recording the item reference. This enables the Rate to be easily found when revisiting the dimensions. For example, D20.014/2 indicates Work Section D20, Item 014, Column 2 in the PSA Schedule of Rates for Building Work.

Column 6 is the "unit" column. This should relate to both the unit of measurement in the Schedule of Rates and the number of dimensions.

Column 7 is the "quantity" column and is transferred from Column 3.

Column 8 is the "rate" column and is the Rate in the PSA Schedule of Rates.

Column 9 is the "extension" column and is Column 7 multiplied by Column 8.

8 Measurement and Valuation

Once the work has been measured and the quantities established it is valued in accordance with the contract. The priority for the valuation of measured work is:

- at schedule rates
- at pro-rata schedule rates
- at fair rates

Schedule rates

These are the published rates contained within the contract schedule of rates and any addendum schedule(s) issued at the time of tender. The values computed using these rates are subject to updating and contract percentage adjustments contained within the contract.

Pro-rata rates

These are rates which are not covered by the pre-priced schedule but are DEDUCED therefrom, e.g. If the schedule contained rates for 50mm and 100mm valves it would be possible to derive a rate for 75mm valves.

The values computed using these rates are also subject to updating and contract percentage adjustments contained within the contract.

Fair rates

Where it is not possible to use schedule or pro-rata rates, fair rates may be used to value the quantities.

It is normal that these rates reflect current costs and therefore are not subject to the contract percentage adjustment or updating percentage.

Where work cannot be measured and valued in accordance with any of the three above, the contract will normally allow for valuation by DAYWORK.

Dayworks

Here, valuation is made on the basis of the labour hours spent multiplied by the operative's hourly rate and the net cost of materials plus a percentage for overheads and profit. The contract will normally determine the labour rates or the method by which it is calculated.

Dayworks provide no incentive to the contractor to work efficiently or seek best value for money when buying materials, etc. Indeed, the opposite is often true; the longer he takes and the more the materials cost, the more he earns!

Sub-contractors and suppliers

Most contracts cater for the employment of sub-contractors and suppliers nominated by the employer. The PSA Schedules of Rates specify how a contractor will be paid in respect of sub-contractors and suppliers, unless the conditions of contract state otherwise.

Overtime and nightworking

Where overtime and nightworking may be required, the contract should define the way in which the contractor will be recompensed for non-productive time. Care should be taken in the administration of contracts that overtime payments are only made where overtime or nightworking has been specifically ordered.

Invoices and discounts

Inclusion of payments based on invoices are dealt with under section A (General Directions) of the PSA Schedule of Rates. Care should be taken to ensure that all discounts and allowances (save 2½% cash discount) are deducted. Care should also be taken to ensure that contractors are not entitled to annual rebates from suppliers. Such rebates will not be obvious from the invoice. Care should also be taken with the use of photo-copy invoices to ensure that they are true copies of the originals and that they accurately reflect the extent of the work executed, i.e. beware of using prices from invoices which contain only small quantities related to the quantity claimed.

Updating and contract percentage adjustments

The application of updating percentages will depend on what is stated in the contract. Some contracts are fixed price for their duration, others are updated annually and some have monthly updating percentages applied. In this case it is normal for the date of the order to determine the month for which percentage will be applied.

Contract percentage adjustments are inserted by the tenderers, and form the basis for tender evaluation.

Normally, on Mechanical and Electrical Services contracts, separate percentages are invited for Rates A, Rates B (including D & E) and Rates C, although further sub-division can occur by inviting separate percentages for each section of the schedule.

9 Worked Examples

The following worked examples have been included. These are based on the PSA Schedules of Rates for Building Work 2009, the PSA Schedules of Rates for Electrical Services 2006 and the PSA Schedules of Rates for Mechanical Services 2006, but the basic principles apply to other editions.

Example BW1: Cutting opening for door in one brick wall
Example BW2: Brick up opening in existing wall
Example BW3: Partition of 100 mm thick blockwork
Example BW4: Repairing defective softwood flooring
Example BW5: Timber stud partition with plasterboard
Example BW6: Door frame
Example BW7: Repairing plaster and lath ceiling
Example BW8: Repairing plaster to concrete ceiling
Example BW9: Patch repair plaster to brick wall
Example BW10: Repairing plaster to brick wall
Example BW11: Repairing wall tiling
Example BW12: Repairing bitumen macadam
Example BW13: Repairing clay roof tiles, battens and felt
Example BW14: Patch repair three layer roofing felt
Example BW15: Re-roofing with three layer roofing felt
Example BW16: Replace sink
Example BW17: Replace basin
Example BW18: Replace WC
Example BW19: Cold water storage tank
Example BW20: Replace PVC gutter and downpipe
Example BW21: Replace length of defective drain pipework
Example BW22: Manhole
Example BW23: Strip foundation (common errors)
Example BW24: Excavation for pit
Example BW25: Suspended reinforced concrete slab
Example E1: Replace fluorescent lamps
Example E2: Renew batten luminaire
Example E3: Renew modular fluorescent luminaire
Example E4: Replace bathroom fan
Example E5: Replace cable fixed to surface
Example E6: Replace cable in conduit
Example E7: Socket outlet on existing main
Example E8: Electrical installation
Example M1: Extending existing copper tubing
Example M2: Inserting fitting into existing copper tubing
Example M3: Replacing copper tubing and stop valve
Example M4: Replacing copper tubing and re-using existing stop valve
Example M5: Pipework renewal
Example M6: Ductwork renewal

Measured Term Contract Order No:- *EXAMPLE BW1*
Location:-
Work to be completed by:-

Please execute the following:-

Cut opening for door in one brick wall, plaster both sides, including precast concrete lintol over and granolithic paving in opening

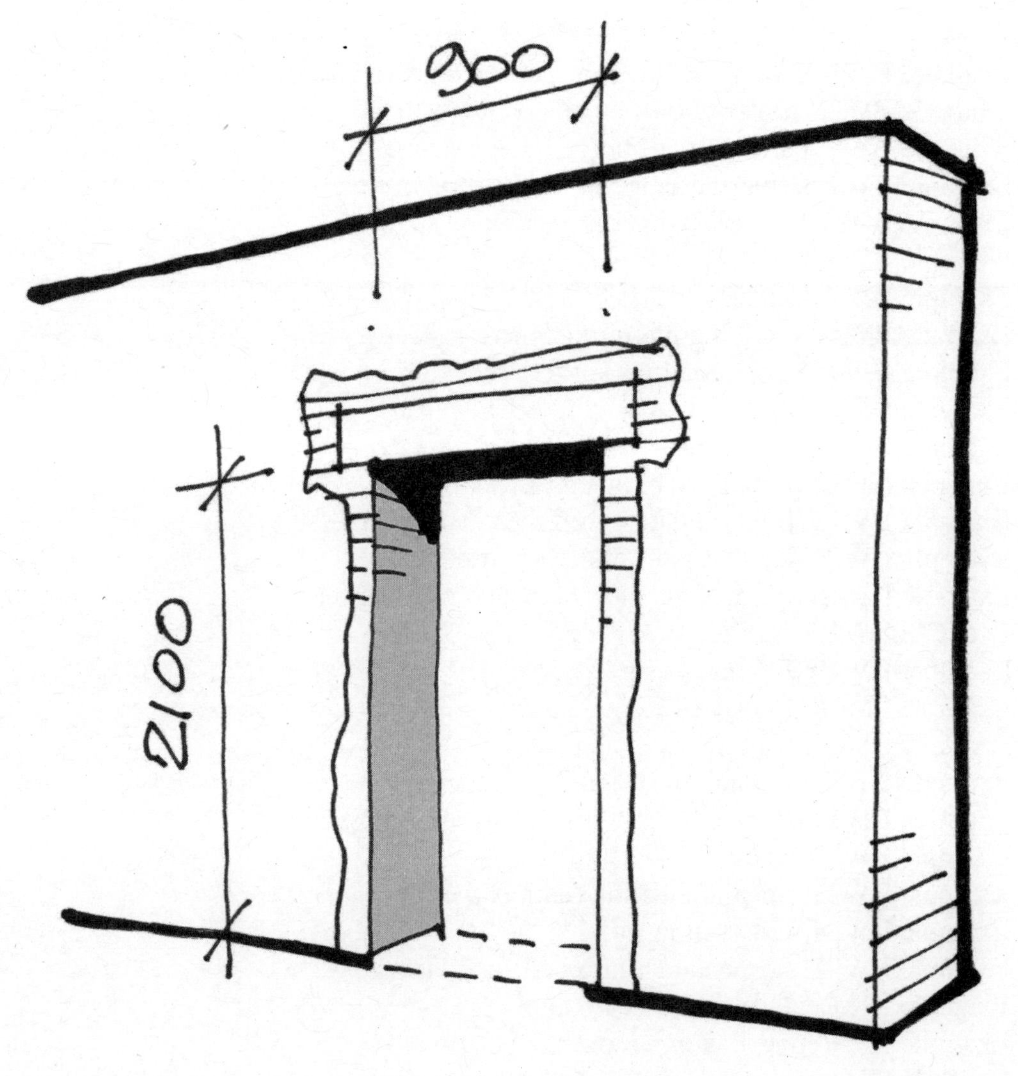

Signed for & on behalf of the PM

....

Date

....

Measured Term Contract Order No: *EXAMPLE BW1*

BUILDING 2009

Measurement & Valuation

				Item ref	Unit	Quantity	Rate	Extension
	1.20	1.20	Precast conc. lintol, 25N/mm², 20mm agg, keyed face	F31.003/3	m	1.20	31.70	38.04
	1.05	1.05	6mm m.s. bar reinforcement 1.05m x 0.222kg/m=0.233kg	E30.010/1	kg	0.23	1.83	0.42
	0.90 2.10 1.20 0.15	 1.89 0.18 2.07	Cutting to form opening through brickwork 2.07m² x 225/25mm = 18.63m²	C90.019	m²	18.63	4.45	82.90
	1.20	1.20	Wedge & pin wi new bkwk over lintol	F30.032/2	m	1.20	21.60	25.92
2	2.10	4.20	Make good jamb of opening	F10.056/2	m	4.20	39.73	166.87
 2	0.90 2.10	0.90 4.20 5.10	Gypsum plaster, wall ne 300mm wide	M20.011/3	m	5.10	9.34	47.63
4 2	2.30 1.20	9.20 2.40 11.60	Jointing to existing plaster	M20.071	m	11.60	4.10	47.56
	0.90 0.23	 0.21	Grano, 40mm th, bonded to conc, fin manually	M10.104/4	m²	0.21	20.96	4.40
			&					
			ne 300mm wide	M10.110/4	m²	0.21	8.70	1.83
			&					
			Anti-slip treatment	M10.155/3	m²	0.21	3.84	0.81
2	0.90	1.80	Jointing to existing grano	M10.145	m	1.80	2.46	4.43
								420.81
			Contract % adjustment			***-10.00%***		-42.08
								378.73
			Add updating % June 2009			0.5%		1.89
			Net total					***£380.62***

Measured Term Contract Order No:- *EXAMPLE BW2*

Location:-

Work to be completed by:-

Please execute the following:-

Brick up opening, one side in facings pointed, other side in commons, plastered, cut, tooth, bond and make good.

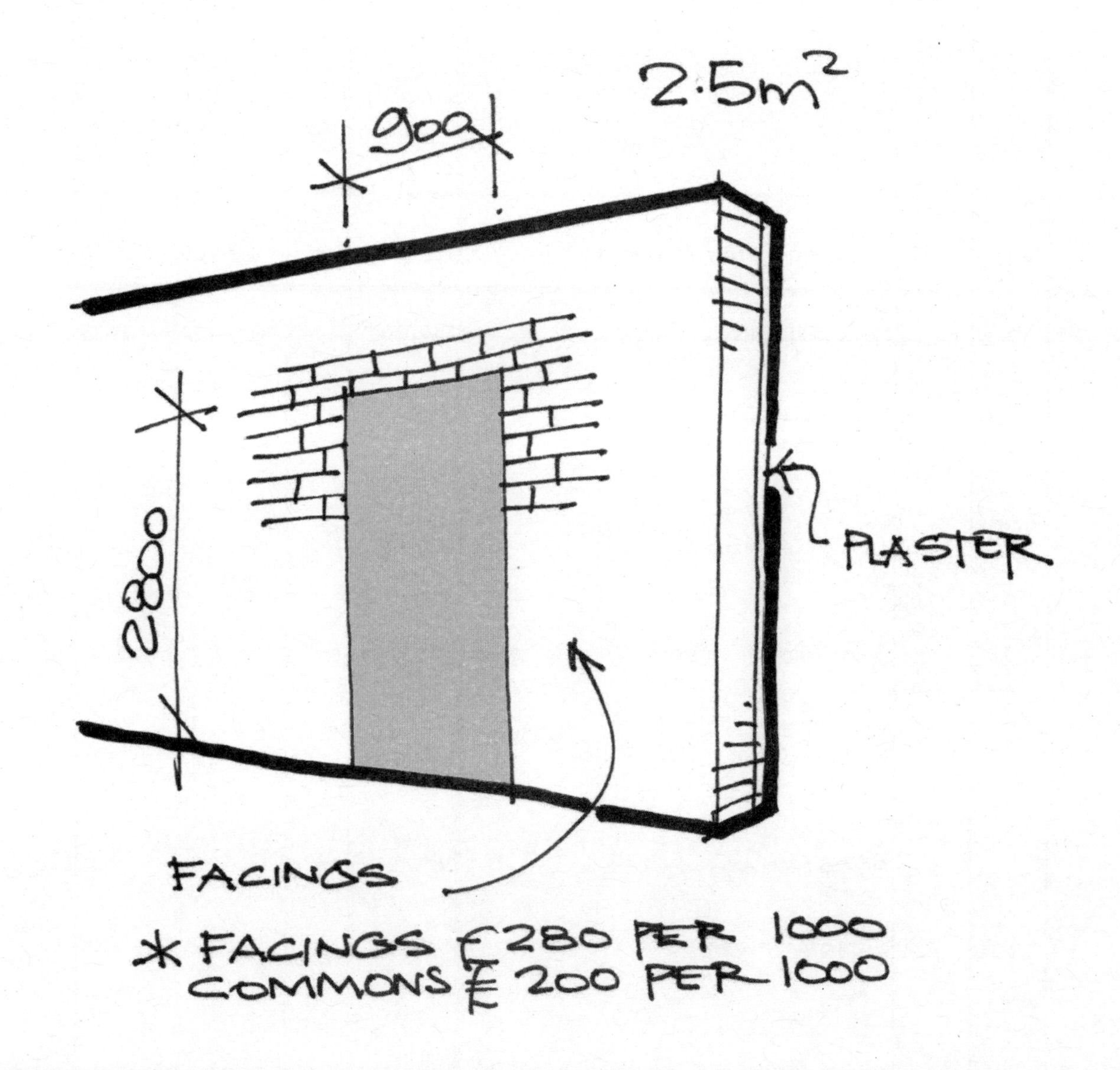

Signed for & on behalf of the PM

....

Date

....

Measured Term Contract Order No: *EXAMPLE BW2*

BUILDING 2009

Measurement & Valuation

				Item ref	Unit	Quantity	Rate	Extension
2	2.80	5.60	Bonding end 1B wall	F10.053/2	m	5.60	22.73	127.29
			&					
			Make good old facings	F10.053/5	m	5.60	3.97	22.23
	0.90	0.90	Wedge & pin 1B wall	F30.031/2	m	0.90	18.60	16.74
			&					
			Make good old facings	F30.031/5	m	0.90	1.32	1.19
	0.90		Wall 1B commons	F10.001/2	m²	2.52	72.47	182.62
	2.80	2.52						
			&					
			Extra over for facings	F10.152/1	m²	2.52	22.66	57.10
			&					
			Fill opening 1B commons	F10.003/2	m²	2.52	25.34	63.86
	0.10		Hack off plaster	C90.102	m²	0.65	3.91	2.54
	0.90	0.09						
2	0.10							
	2.80	0.56						
		0.65						
	0.90		Plaster wall	M20.001/3	m²	2.52	14.02	35.33
	2.80	2.52						
	0.90	0.90	Fair joint	M20.071	m	6.50	4.10	26.65
2	2.80	5.60						
		6.50						
								535.55
			Contract % adjustment			***-10.00%***		-53.56
								481.99
			Add updating % June 2009			0.5%		2.41
								484.40
			ADD bricks (as per invoice)					
			2.52m² x 128-81 = 119		Thou	***0.119***	***200.00***	23.80
			2.52m² x 81 = 205		Thou	***0.205***	***280.00***	57.40
								81.20
					Profit		***5.00%***	4.06
								85.26
			Net total					***£569.66***

Measured Term Contract Order No:-	*EXAMPLE BW3*
Location:-	
Work to be completed by:-	

Please execute the following:-

100 mm thick blockwork in partitions, plastered both sides, with 19 x 75 mm skirting, including cutting and bonding, reinforced PC lintol over openings, cutting back and making good. 2 coats emulsion paint on walls and 2 coats oil paint on skirtings.

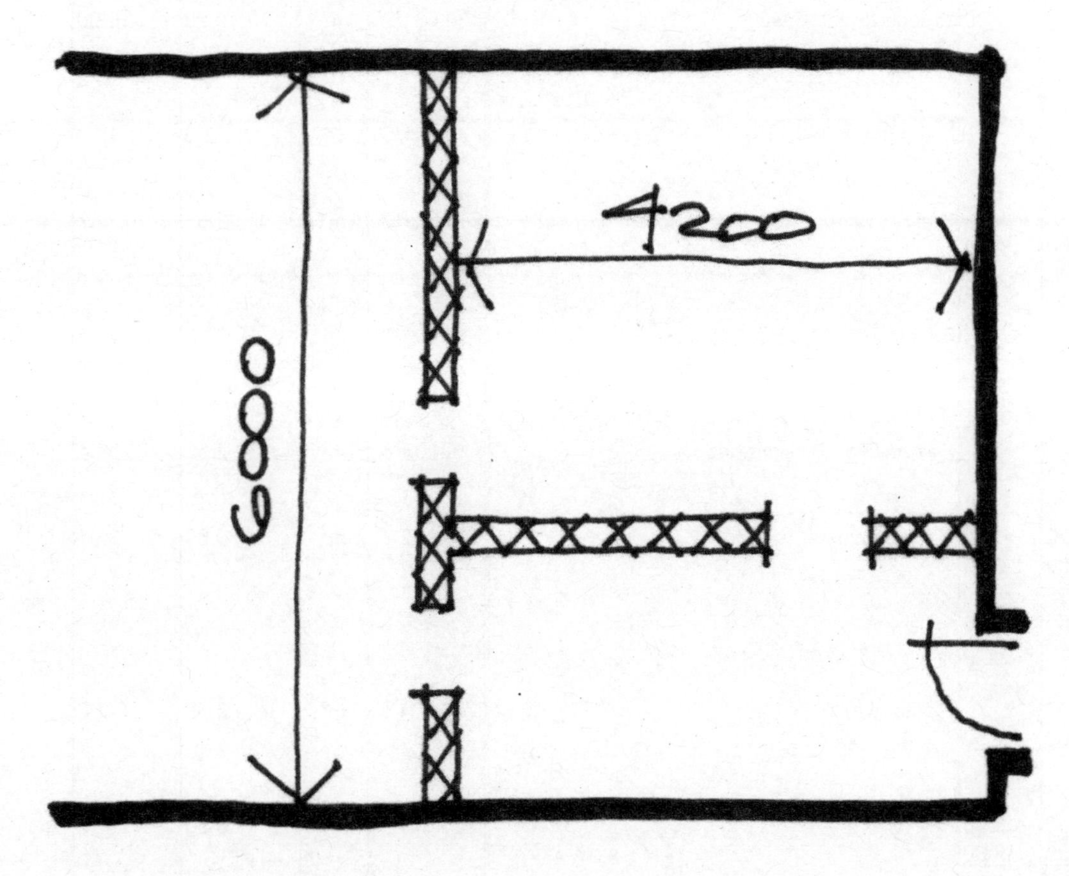

Signed for & on behalf of the PM

....

Date

....

Measured Term Contract Order No: *EXAMPLE BW3*

BUILDING 2009

Measurement & Valuation

				Item ref	*Unit*	*Quantity*	*Rate*	*Extension*
6	1.15	6.90	6mm m.s. bar reinforcement	E30.010/1	kg	1.53	1.83	2.80
			6.9m x 0.222kg/m=1.53kg					
3	1.20	3.60	Lintel, rect., keyed 20N/mm², 10mm agg.	F31.002/2	m	3.60	28.54	102.74
	6.00		Wall in 100mm aerated conc blocks	F10.201/2	m²	18.81	39.74	747.51
	2.40	14.40						
	4.20							
	2.40	10.08						
		24.48						
3	0.90		Ddt for openings					
	2.10	5.67						
		18.81						
3	2.40	7.20	Bonding end new wall to extg	F10.223/1	m	7.20	6.41	46.15
	6.00	6.00	Wedge & pin wi new blkwk over lintol	F30.034/1	m	10.20	8.40	85.68
	4.20	4.20						
		10.20						
	6.00	6.00	19 x 75mm (fin) sw skrting	P20.014/1	m	14.90	2.97	44.25
	5.90	5.90	size 19 x 75 = 1425mm²	P20.014/2	m	14.90	3.05	45.45
2	4.20	8.40	P20.014/2					
		20.30	(1425 - 500)/100 = 9.25 x 0.33 = £3.05					
6	0.90	5.40	Ddt for openings					
		14.90	&					
			Paint 1 ct prime 1 u/ct 1 fin ct oil paint isol surf ne 300mm g	M60.231/1	m	14.90	3.71	55.28
	6.00		Gypsum plaster, wall ne 300mm wide	M20.001/3	m²	38.91	14.02	545.52
	2.40	14.40	2 cts on blkwk					
	5.90							
	2.40	14.16						
2	4.20							
	2.40	20.16						
3	0.90		(around					
	0.10	0.27	(door					
6	2.10		(opening					
	0.10	1.26						
		50.25						
6	0.90		Ddt for openings					
	2.10	11.34						
		38.91	&					
			Paint 1 ct matt emul plaster wall	M60.026/1	m²	38.91	2.44	94.94
			Additional coat	M60.026/3	m²	38.91	1.69	65.76
								1836.08
			Contract % adjustment			***-10.00%***		-183.61
								1652.47
			Add updating % June 2009			0.5%		8.26
			Net total					***£1,660.73***

Measured Term Contract Order No:- *EXAMPLE BW4*

Location:-

Work to be completed by:-

Please execute the following:-

Take up and renew defective softwood flooring joists and wallplate

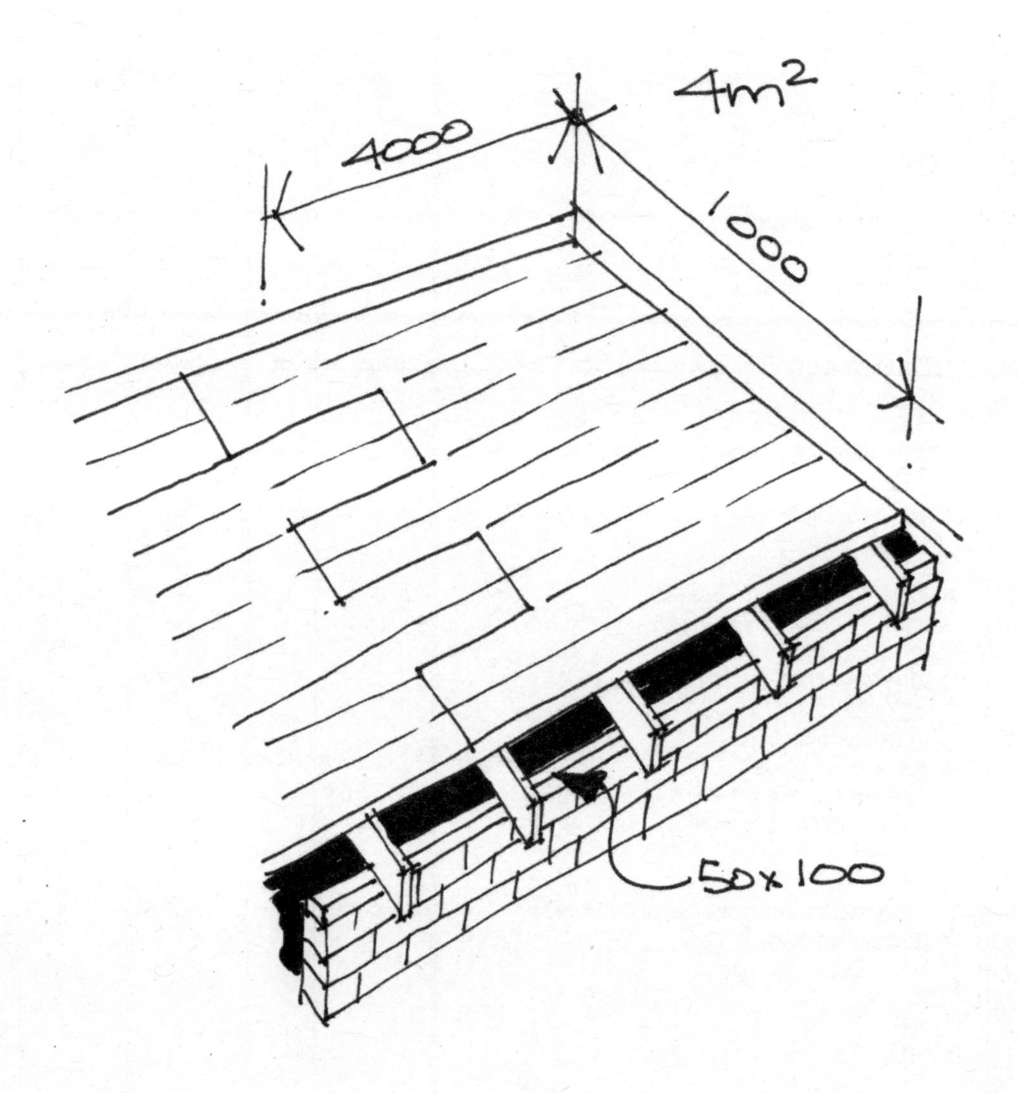

Signed for & on behalf of the PM

....

Date

....

Measured Term Contract Order No: *EXAMPLE BW4*

BUILDING 2009

Measurement & Valuation

				Item ref	Unit	Quantity	Rate	Extension
			nr of joists @ 350 mm c/c 4000 / 350 = 11 spaces = 10 joists					
10	1.00	10.00	Take down timber 50 x 100 = 5000 mm² £3.08 / 10000 x 5000 = £1.54	C90.035	m	10.00	1.54	15.40
10	1.10	11.00	Preserve timber 50 x 100 = 5000 mm² £0.40 / 10000 x 5000 = £0.20	G20.001	m	11.00	0.20	2.20
10	1.10	11.00	Floor member 50 x 100 = 5000 mm² £8.16 / 10000 x 5000 = £4.08	G20.005/1	m	11.00	4.08	44.88
	4.00 1.00	4.00	Take up flooring	C90.068/3	m²	4.00	5.94	23.76
			&					
			Wrot sw flooring 21mm th	K20.031/3	m²	4.00	23.53	94.12
			&					
			In repairs 1 to 5m²	K20.048/3	m²	4.00	15.59	62.36
	1.00	1.00	Jointing to extg	K20.057	m	1.00	4.10	4.10
								246.82
			Contract % adjustment			***-10.00%***		-24.68
								222.14
			Add updating % June 2009			0.5%		1.11
			Net total					***£223.25***

Measured Term Contract Order No:- *EXAMPLE BW5*

Location:-

Work to be completed by:-

Please execute the following:-

Timber stud partition, lined both sides with 9.5 mm plasterboard and skim coat gypsum plaster, including 19 x 75 mm skirting and 20 x 20 mm quadrant moulding. 2 coats emulsion paint on walls and 2 coats oil paint on skirtings.

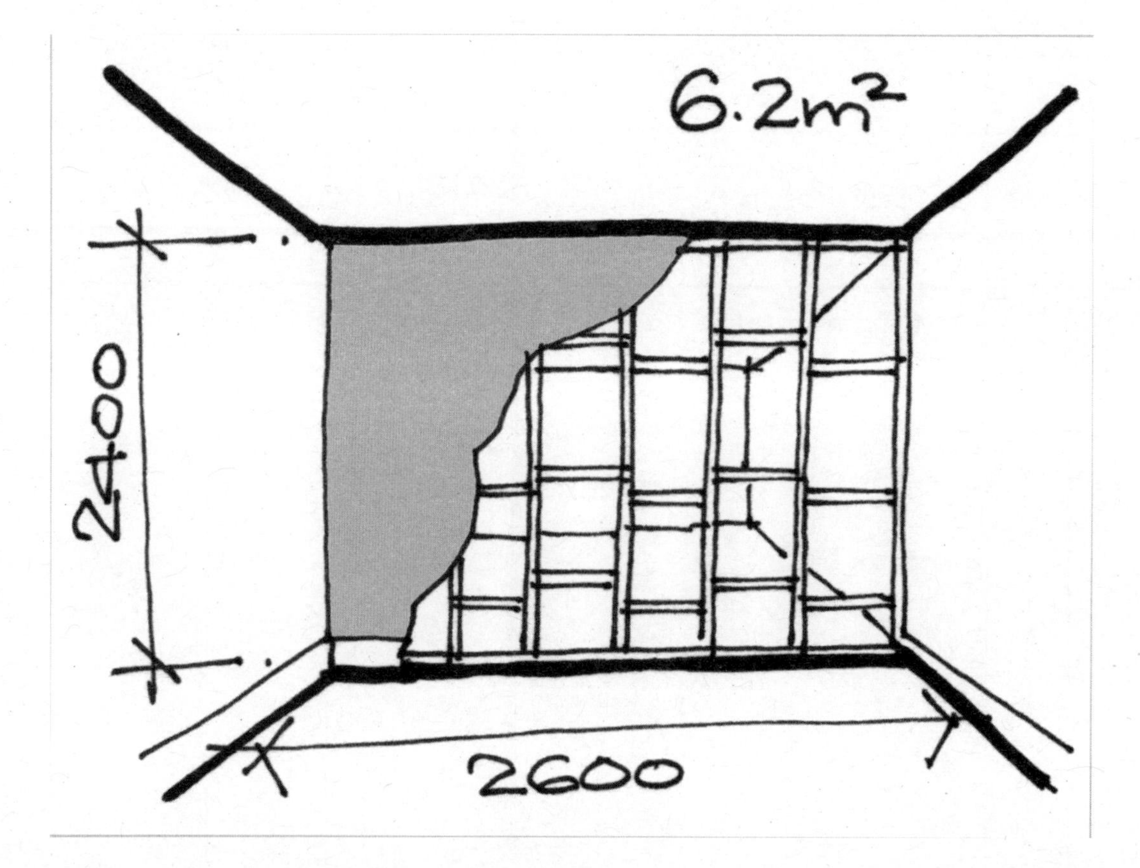

Signed for & on behalf of the PM

....

Date

....

Measured Term Contract Order No: *EXAMPLE BW5*

BUILDING 2009

Measurement & Valuation

				Item ref	*Unit*	*Quantity*	*Rate*	*Extension*
5	2.60	13.00	Partition member	G20.006/1	m	29.80	4.965	147.96
7	2.40	16.80	50 x 100 = 5000 mm²					
		29.80	£9.93 / 10000 x 5000 = £4.965					
2	2.60	5.20	Quadrant 20 x 20 mm	P20.010	m	14.52	2.25	32.67
4	2.33	9.32	2.600 2.400					
		14.52	Ddt sktg -0.075					
			2.600 2.325					
2	2.60	5.20	19 x 75mm (fin) sw skrting	P20.014/1	m	5.20	2.97	15.44
			size 19 x 75 = 1425mm²	P20.014/2	m	5.20	3.05	15.86
			P20.014/2					
			(1425 - 500)/100 = 9.25 x 0.33 = £3.05					
			&					
			Labour on skirting	P20.031/1	m	5.20	0.54	2.81
			&					
			Paint 1 ct prime 1 u/ct 1 fin ct oil paint isol surf ne 300mm g	M60.231/1	m	5.20	3.71	19.29
4	2.50	10.00	Screws	P20.115/1	m	10.00	0.68	6.80
			&					
			Plugging	P20.119	m	10.00	1.68	16.80
2	2.60		9.5mm baseboard to wall	M20.086/1	m²	12.48	8.99	112.20
	2.40	12.48						
			&					
			Gypsum plaster, wall exc 300mm wide 1 ct on plasterboard	M20.001/2	m²	12.48	7.94	99.09
			&					
			Paint 1 ct matt emul plaster wall	M60.026/1	m²	12.48	2.44	30.45
			Additional coat	M60.026/3	m²	12.48	1.69	21.09
								520.46
			Contract % adjustment			***-10.00%***		-52.05
								468.41
			Add updating % June 2009			0.5%		2.34
			Net total					***£470.75***

Measured Term Contract Order No:- *EXAMPLE BW6*

Location:-

Work to be completed by:-

Please execute the following:-

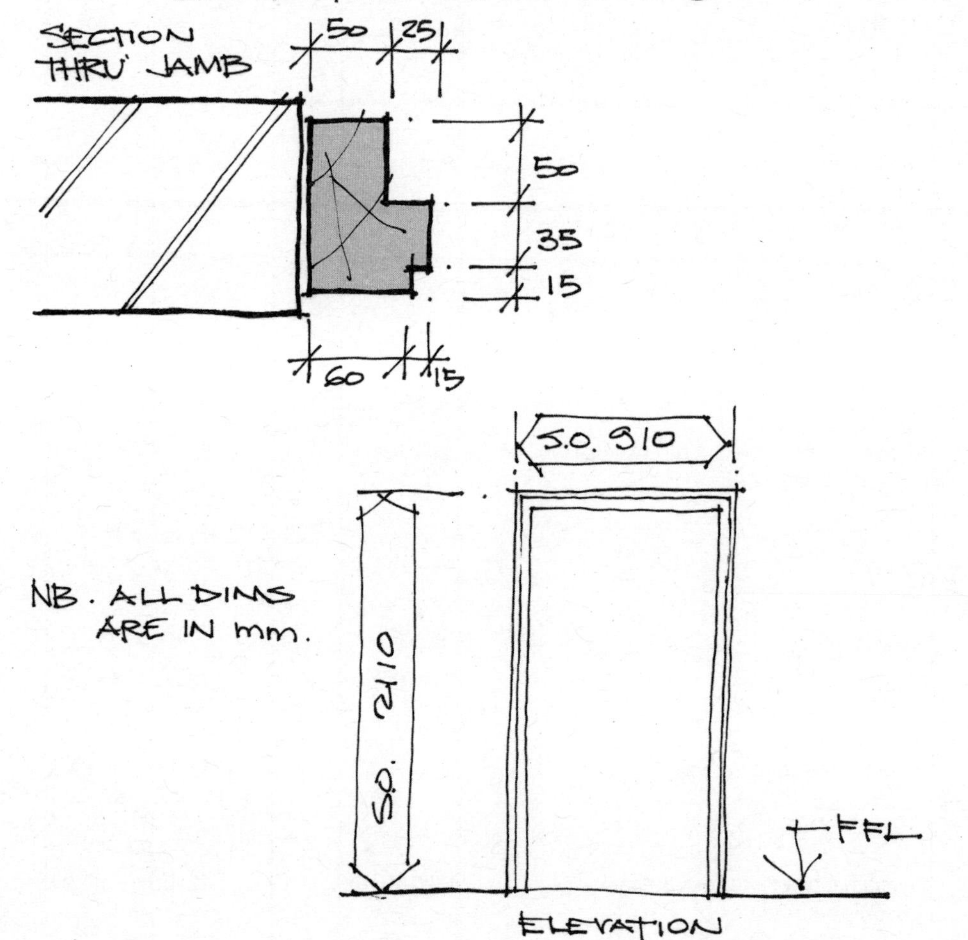

Signed for & on behalf of the PM

....

Date

....

Measured Term Contract Order No:- *EXAMPLE BW6*

BUILDING 2009

Measurement & Valuation

				Item ref	*Unit*	*Quantity*	*Rate*	*Extension*
			Frame					
	0.90	0.90	Dr fr/ hw/ 100 x 75mm (nom)/ teak	L20.103/3	m		2.09	
2	2.10	4.20		L20.103/4	m		16.20	
		5.10					18.29	
				L10.001	factor		1.05	
			Finished size 94 x 69mm = 6486mm²	K20.007	factor		2.50	
						5.10	48.01	244.86
			L20.103/4					
			(6486 - 500)/100 = 60 x 0.27 = £16.20					
			Labours					
2	5.10	10.20	Rebate	L20.110/2	m		1.10	
				K20.007	factor		2.50	
						10.20	2.75	28.05
			Fixing					
	5.10	5.10	Plugging	P20.119	m	5.10	1.68	8.57
			&					
			Screwing & pelleting	P20.116/2	m		1.41	
				P20.118/2	m		4.29	
							5.70	
				K20.007	factor		2.50	
						5.10	14.25	72.68
								354.16
			Contract % adjustment			***-10.00%***		-35.42
								318.74
			Add updating % June 2009			0.5%		1.59
			Net total					***£320.33***

Measured Term Contract Order No:- *EXAMPLE BW7*

Location:-

Work to be completed by:-

Please execute the following:-

Cut out and take down old plaster and lath ceiling and renew with 9.5 mm gypsum plasterboard and skim coat plaster.

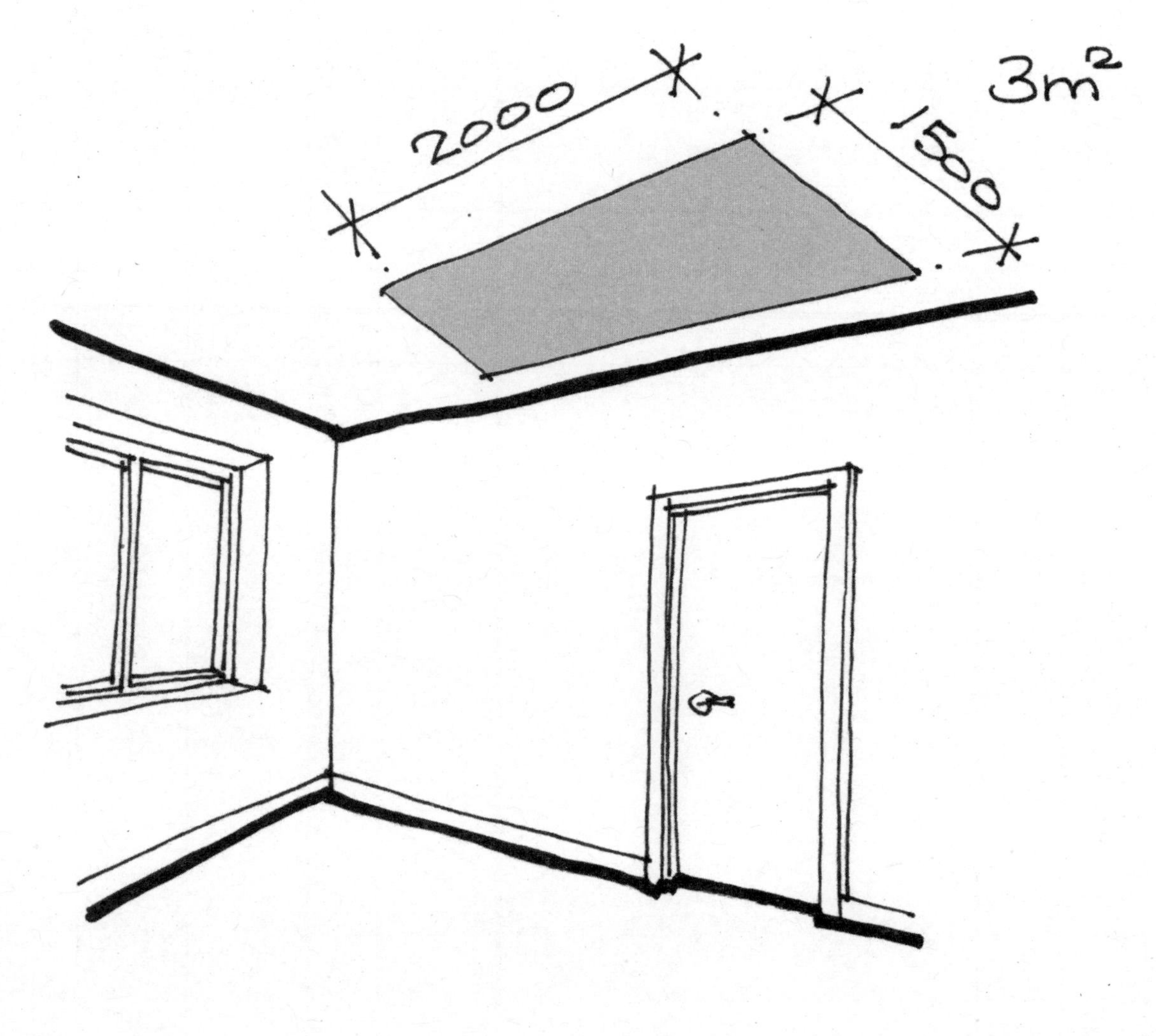

Signed for & on behalf of the PM

....

Date

....

Measured Term Contract Order No: *EXAMPLE BW7*

BUILDING 2009

Measurement & Valuation

				Item ref	*Unit*	*Quantity*	*Rate*	*Extension*
	2.00 1.50	3.00	Gypsum plaster, ceiling exc 300mm wide 1 ct on plasterboard	M20.002/2	m²	3.00	8.60	25.80
			&					
			In repairs 1 to 5m²	M20.008/2	m²	3.00	9.76	29.28
			&					
			9.5mm baseboard to ceiling	M20.087/1	m²	3.00	9.57	28.71
			&					
			In repairs 1 to 5m²	M20.095/1	m²	3.00	11.63	34.89
								118.68
			Contract % adjustment			***-10.00%***		-11.87
								106.81
			Add updating % June 2009			0.5%		0.53
			Net total					***£107.34***

Measured Term Contract Order No:- *EXAMPLE BW8*

Location:-

Work to be completed by:-

Please execute the following:-

Hack off old plaster from concrete ceiling and renew with gypsum plaster in 2 coats including dubbing out up to 5 mm as necessary

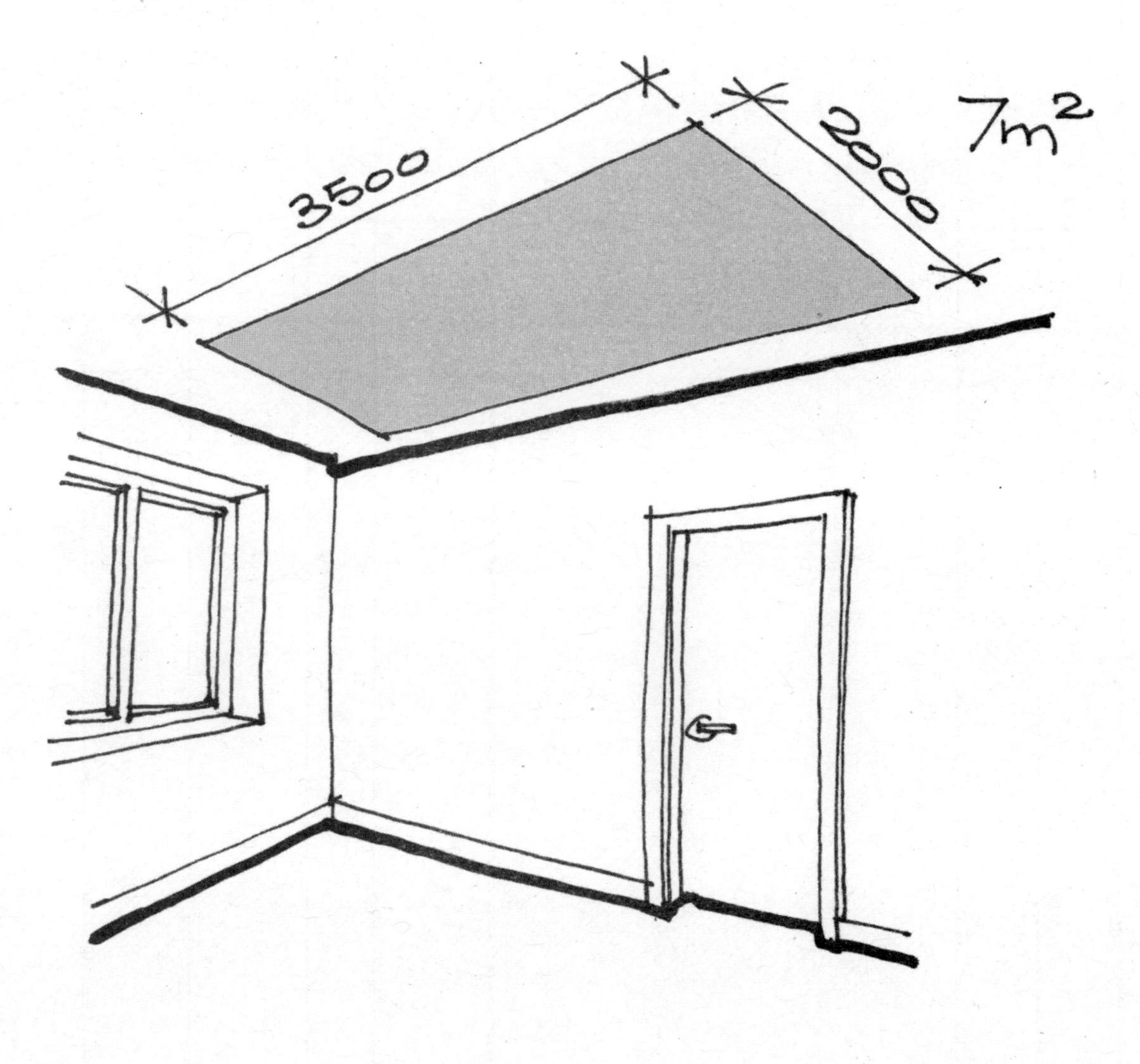

Signed for & on behalf of the PM

....

Date

....

Measured Term Contract Order No: *EXAMPLE BW8*

BUILDING 2009

Measurement & Valuation

				Item ref	*Unit*	*Quantity*	*Rate*	*Extension*
	3.50 2.00	7.00	Hack off plaster from ceiling	C90.103	m²	7.00	4.73	33.11
			&					
			Gypsum plaster, ceiling exc 300mm wide 2 cts on concrete	M20.002/4	m²	7.00	17.16	120.12
			&					
			Dubbing out, exc 300mm wide, extg backgnd ne 5mm av th	M20.009/4	m²	7.00	7.11	49.77
2 2	3.50 2.00	7.00 4.00 11.00	Joint to existing	M20.071	m	11.00	4.10	45.10
								248.10
			Contract % adjustment			**-10.00%**		-24.81
								223.29
			Add updating % June 2009			0.5%		1.12
			Net total					***£224.41***

Measured Term Contract Order No:- *EXAMPLE BW9*

Location:-

Work to be completed by:-

Please execute the following:-

Hack off old plaster from wall, form key, and renew with gypsum plaster in 2 coats including dubbing out up to 5 mm as necessary

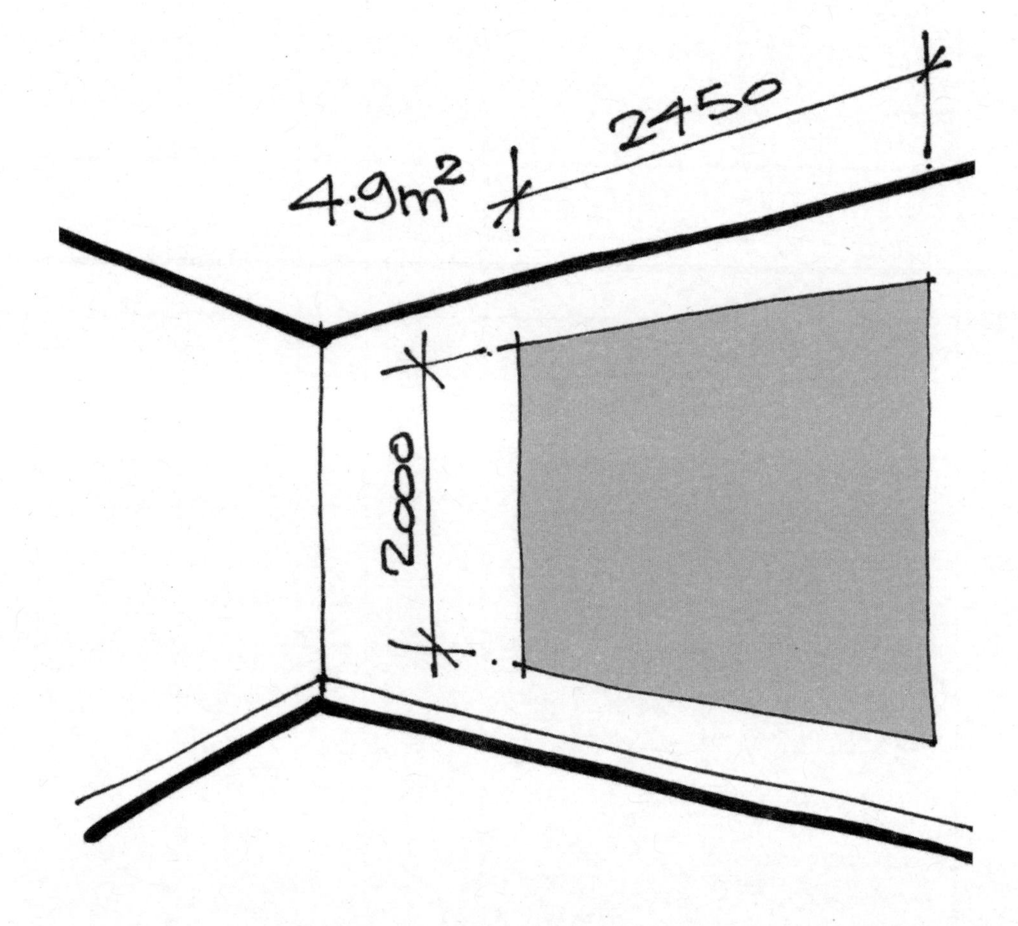

Signed for & on behalf of the PM

....

Date

....

Measured Term Contract Order No: *EXAMPLE BW9*

BUILDING 2009

Measurement & Valuation

				Item ref	Unit	Quantity	Rate	Extension
	2.45 2.00	4.90	Gypsum plaster, wall exc 300mm wide 2 cts on brickwork	M20.001/3	m²	4.90	14.02	68.70
			&					
			In repairs 1 to 5m²	M20.008/3	m²	4.90	18.86	92.41
								161.11
			Contract % adjustment			***-10.00%***		-16.11
								145.00
			Add updating % June 2009			0.5%		0.73
			Net total					***£145.73***

Measured Term Contract Order No:- *EXAMPLE BW10*

Location:-

Work to be completed by:-

Please execute the following:-

Hack off old plaster from wall, form key, and renew with gypsum plaster in 2 coats including dubbing out up to 5 mm as necessary

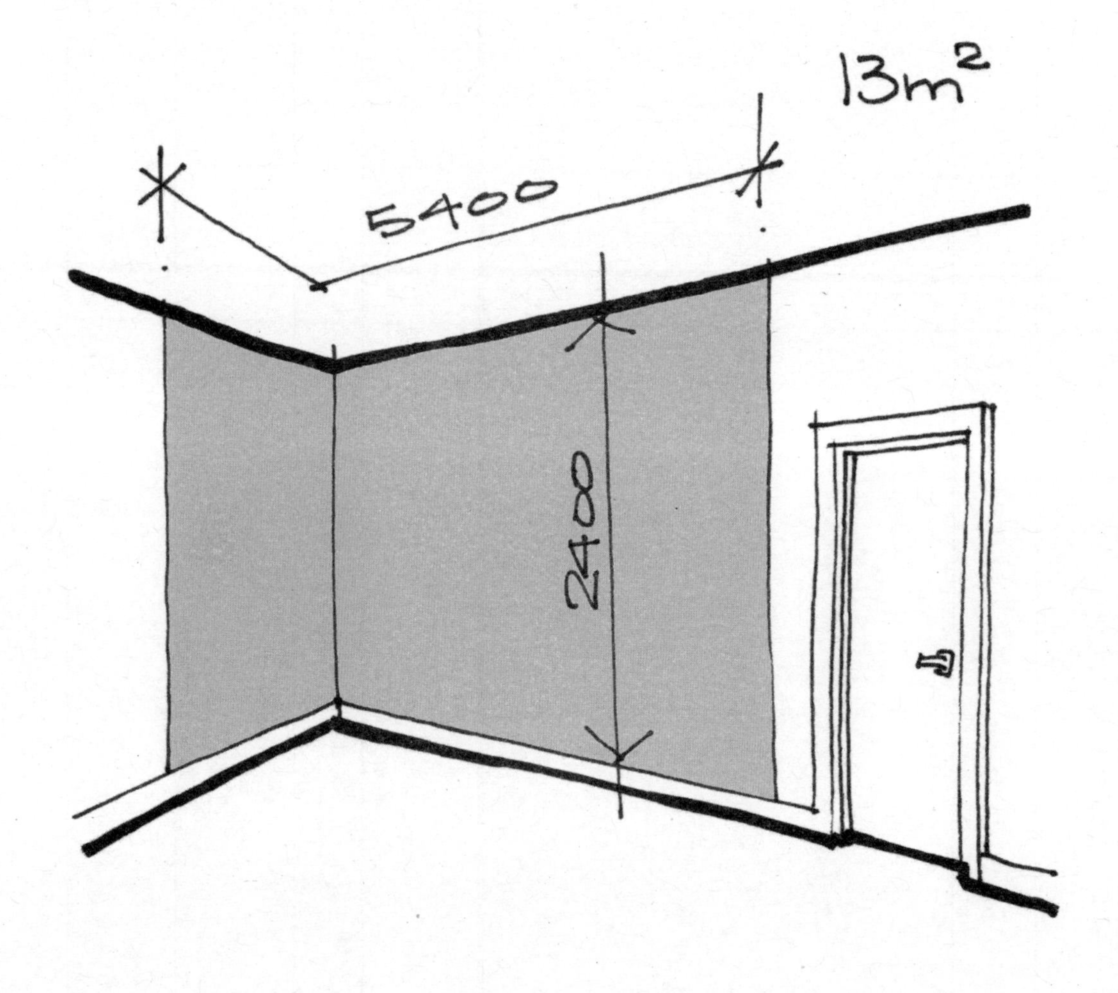

Signed for & on behalf of the PM

....

Date

....

Measured Term Contract Order No: *EXAMPLE BW10*

BUILDING 2009

Measurement & Valuation

				Item ref	*Unit*	*Quantity*	*Rate*	*Extension*
	5.60 2.40	13.44	Hack off plaster from wall	C90.102	m^2	13.44	3.91	52.55
			&					
			Form key	F10.231	m^2	13.44	3.25	43.68
			&					
			Gypsum plaster, wall exc 300mm wide 2 cts on brickwork	M20.001/3	m^2	13.44	14.02	188.43
			&					
			Dubbing out, exc 300mm wide, extg backgnd ne 5mm av th	M20.009/3	m^2	13.44	6.66	89.51
2	2.40	4.80	Joint to existing	M20.071	m	4.80	4.10	19.68
								393.85
			Contract % adjustment			***-10.00%***		-39.39
								354.46
			Add updating % June 2009			0.5%		1.77
			Net total					***£356.23***

Measured Term Contract Order No:- *EXAMPLE BW11*

Location:-

Work to be completed by:-

Please execute the following:-

HACK OFF OLD WALL TILING AND BACKING FROM WALL, FORM KEY, DUB OUT AS NECESSARY, APPLY 15mm CEMENT AND SAND BACKING.
SUPPLY AND FIX 152 x 152 x 5.5 mm WHITE GLAZED WALL TILES

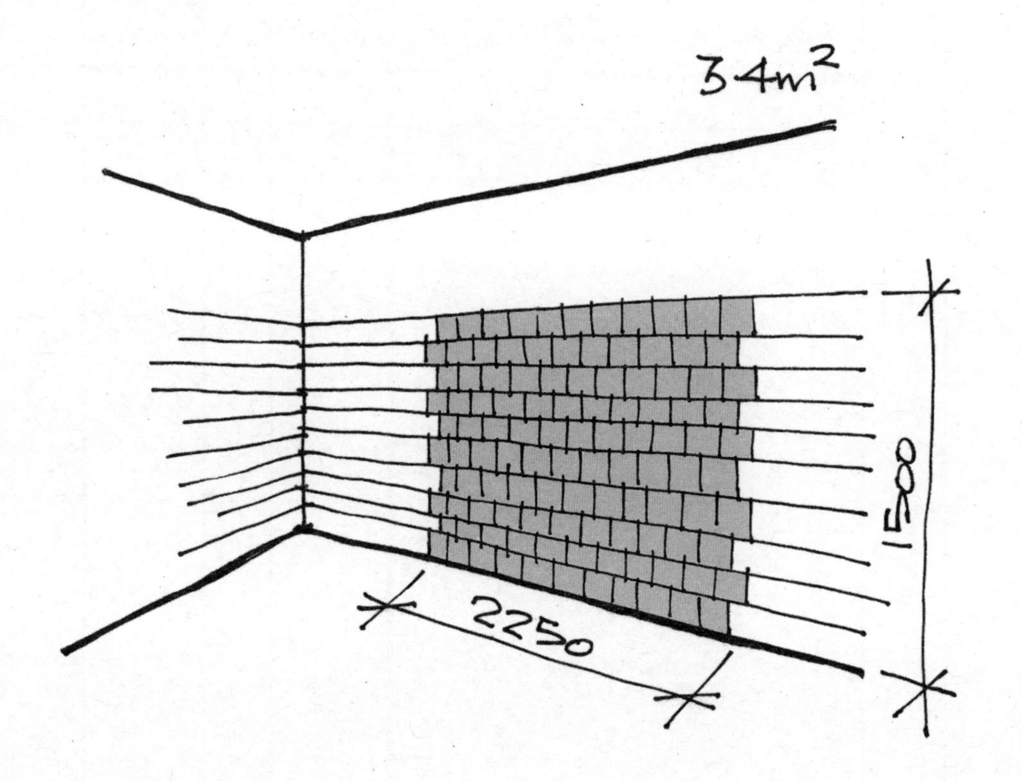

Signed for & on behalf of the PM

....

Date

....

Measured Term Contract Order No: *EXAMPLE BW11*

BUILDING 2009

Measurement & Valuation

				Item ref	*Unit*	*Quantity*	*Rate*	*Extension*
	2.25		Render internal wall exc 300mm wide	M20.044/2	m²	3.38	16.74	56.58
	1.50	3.38	2 cts on brickwork					
			&					
			In repairs 1 to 5m²	M20.051/2	m²	3.38	22.25	75.21
			&					
			152 x 152 x 5.5mm white glazed wall tiles	M40.048/2	m²	3.38	46.03	155.58
			&					
			In repairs 1 to 5m²	M40.055/2	m²	3.38	49.14	166.09
								453.46
			Contract % adjustment			***-10.00%***		-45.35
								408.11
			Add updating % June 2009			0.5%		2.04
			Net total					***£410.15***

Measured Term Contract Order No:- *EXAMPLE BW12*

Location:-

Work to be completed by:-

Please execute the following:-

LAY COATED MACADAM TO EXISTING MACADAM

1. AREAS BETWEEN 10 & 25m²;
2. 500m² IN ONE AREA;
3. 250 PATCHES AT 2m² EACH

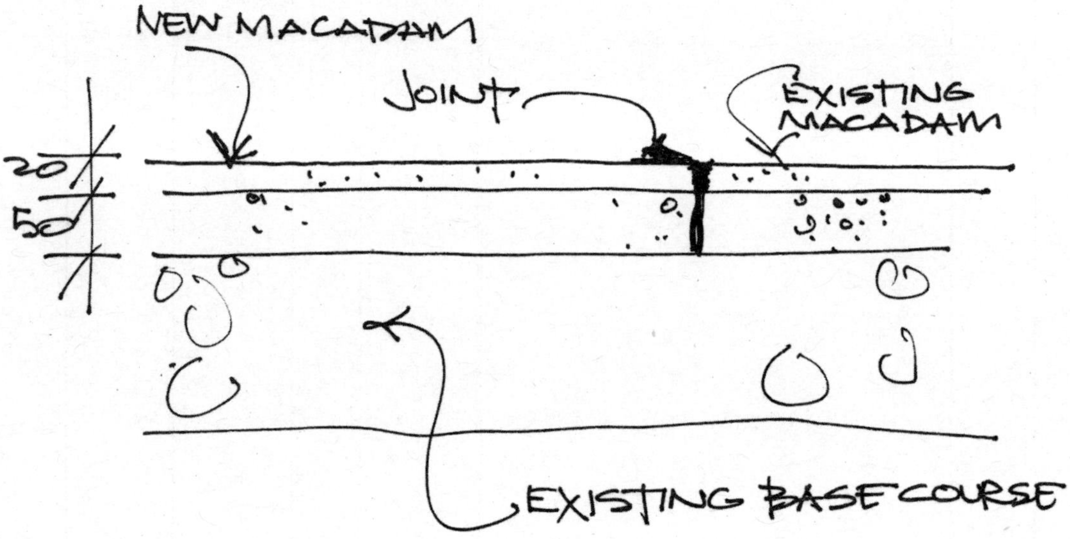

NEW MACADAM

JOINT

EXISTING MACADAM

20

50

EXISTING BASECOURSE

Signed for & on behalf of the PM

....

Date

....

Measured Term Contract Order No:- *EXAMPLE BW12*

BUILDING 2009

Measurement & Valuation

				Item ref	Unit	Quantity	Rate	Extension
			Area between 10 and 25 m²					
	5.00		Base course £11.81-£0.97*2	Q22.004/1	m²	20.00	9.87	197.40
	4.00	20.00	&					
			Wearing course	Q22.008/1	m²	20.00	4.75	95.00
2	5.00	10.00	Jointing to existing	Q22.011	m	18.00	1.20	21.60
2	4.00	8.00						
		18.00						314.00
			Factor	Q22.002	1.3			408.20
								408.20
			Contract % adjustment			*-10.00%*		-40.82
								367.38
			Add updating % June 2009			0.5%		1.84
			Net total					***£369.22***

				Item ref	Unit	Quantity	Rate	Extension
			500 m² in one area					
	50.00		Base course £11.81-£0.97*2	Q22.004/1	m²	500.00	9.87	4935.00
	10.00	500.00	&					
			Wearing course	Q22.008/1	m²	500.00	4.75	2375.00
2	50.00	100.00	Jointing to existing	Q22.011	m	120.00	1.20	144.00
2	10.00	20.00						
		120.00						
								7454.00
			Contract % adjustment			*-10.00%*		-745.40
								6708.60
			Add updating % June 2009			0.5%		33.54
			Net total					***£6,742.14***

				Item ref	Unit	Quantity	Rate	Extension
			250 patches each 2 m²					
250	2.00		Base course £11.81-£0.97*2	Q22.004/1	m²	500.00	9.87	4935.00
	1.00	500.00	&					
			Wearing course	Q22.008/1	m²	500.00	4.75	2375.00
500	2.00	1000.00	Jointing to existing	Q22.011	m	1500.00	1.20	1800.00
500	1.00	500.00						
		1500.00						
								9110.00
			Contract % adjustment			*-10.00%*		-911.00
								8199.00
			Add updating % June 2009			0.5%		41.00
			Net total					***£8,240.00***

Measured Term Contract Order No:- *EXAMPLE BW13*

Location:-

Work to be completed by:-

Please execute the following:-

Take off and renew battens, felt and plain clay tiles

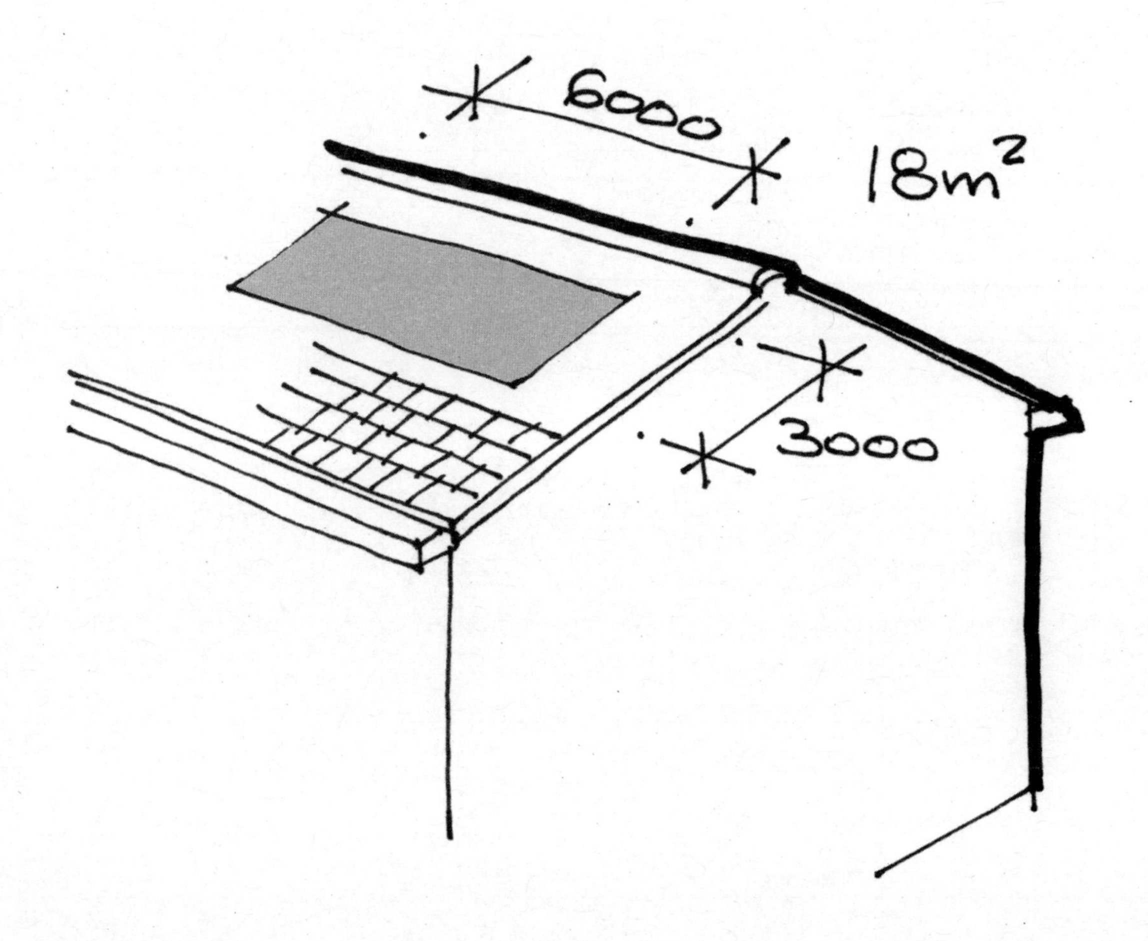

Signed for & on behalf of the PM

....

Date

....

Measured Term Contract Order No: *EXAMPLE BW13*

BUILDING 2009

Measurement & Valuation

				Item ref	Unit	Quantity	Rate	Extension
	6.00 3.00	18.00	Stripping tiling	C90.051	m²	18.00	3.36	60.48
			&					
			Taking off battens	C90.052	m²	18.00	1.21	21.78
			&					
			Stripping felt underlay	C90.054	m²	18.00	1.44	25.92
			&					
			Fixing only plain tiles	H60.001	m²	18.00	21.66	389.88
			&					
			25 x 50mm battens	H60.006	m²	18.00	1.70	30.60
2 2	6.00 3.00	12.00 6.00 18.00	Jointing to existing	H60.039	m	18.00	5.74	103.32
	6.00 3.00	18.00	Felt underlay, fixing wi battens	H60.051/1	m²	18.00	7.57	136.26
								768.24
			Contract % adjustment			***-10.00%***		-76.82
								691.42
			Add updating % June 2009			0.5%		3.46
								694.88
			ADD tiles (as per invoice) ***18m² x 58 = 1044***		Thou	***1.044***	***220.00***	229.68
					Profit		***5.00%***	11.48
			Net total					***£936.04***

Measured Term Contract Order No:- *EXAMPLE BW14*

Location:-

Work to be completed by:-

Please execute the following:-

PATCH REPAIR THREE LAYERS FELT ROOFING CUT OUT THREE LAYER FELT, INSULATION BOARD, LAY NEW INCLUDING NEW INSULATION BOARD, VAPOUR BARRIER, AND CHIPPINGS.

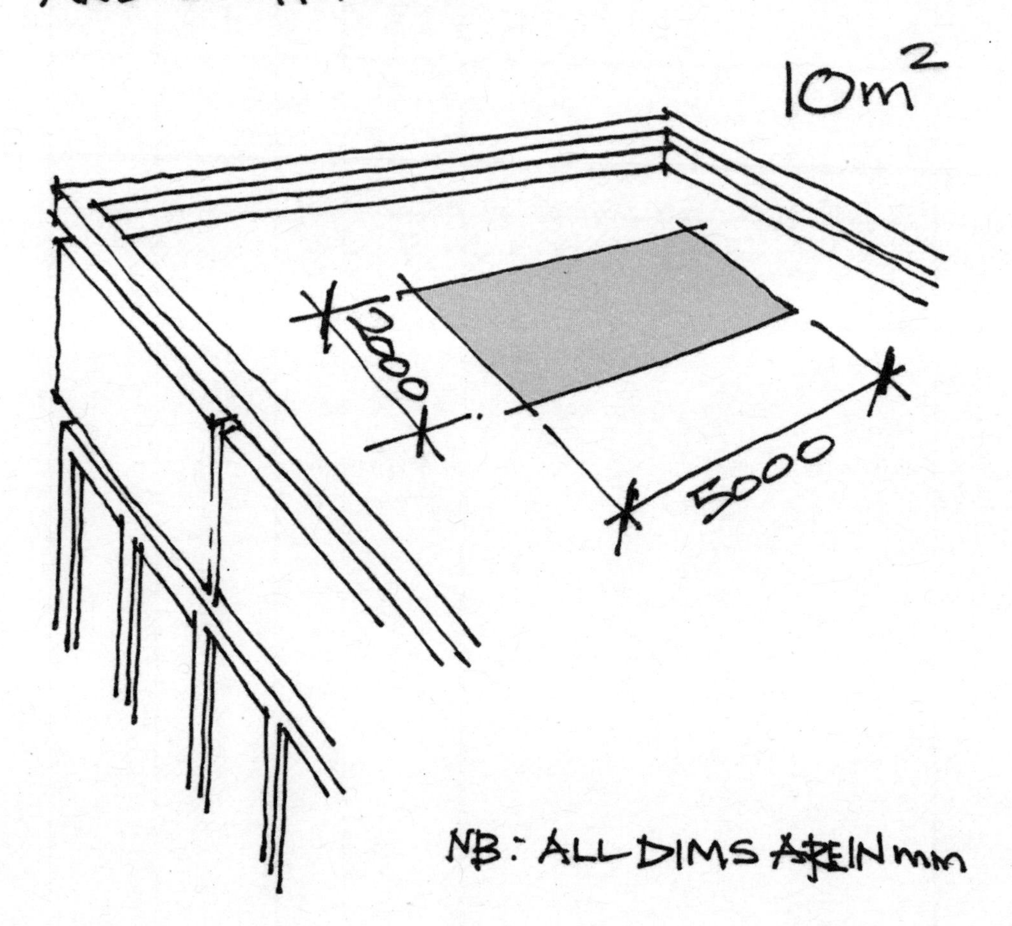

Signed for & on behalf of the PM

....

Date

....

Measured Term Contract Order No: *EXAMPLE BW14*

BUILDING 2009

Measurement & Valuation

				Item ref	*Unit*	*Quantity*	*Rate*	*Extension*
	2.00 5.00	10.00	Stripping felt roofing, to masonry, any number layers	C90.064	m²	10.00	3.13	31.30
			&					
			Elastomeric bottom layer	J41.001/4	m²	10.00	8.24	82.40
			&					
			In repairs 5 to 10m²	J41.010/4	m²	10.00	3.15	31.50
			&					
			Elastomeric intermediate layer	J41.001/4	m²	10.00	8.24	82.40
			&					
			In repairs 5 to 10m²	J41.010/4	m²	10.00	3.15	31.50
			&					
			Elastomeric cap sheet	J41.001/5	m²	10.00	14.29	142.90
			&					
			In repairs 5 to 10m²	J41.010/5	m²	10.00	3.40	34.00
			&					
			Insulation glassboard	J41.037/4	m²	10.00	18.39	183.90
			&					
			In repairs 5 to 10m²	J41.041/4	m²	10.00	2.88	28.80
			&					
			ADD for priming base	J41.005/4	m²	10.00	2.15	21.50
			&					
			Chippings	J41.007/5	m²	10.00	5.78	57.80
			&					
			Cleaning base after removal of felt	J41.046	m²	10.00	0.98	9.80
								737.80
			Contract % adjustment			***-10.00%***		-73.78
								664.02
			Add updating % June 2009			0.5%		3.32
			Net total					***£667.34***

Measured Term Contract Order No:- *EXAMPLE BW15*

Location:-

Work to be completed by:-

Please execute the following:-

CUT OUT THREE LAYERS FELT ROOFING CUT OUT INSULATION BOARD, LAY NEW ELASTOMERIC FELT INCLUDING NEW INSULATION BOARD, VAPOUR BARRIER, AND CHIPPINGS.

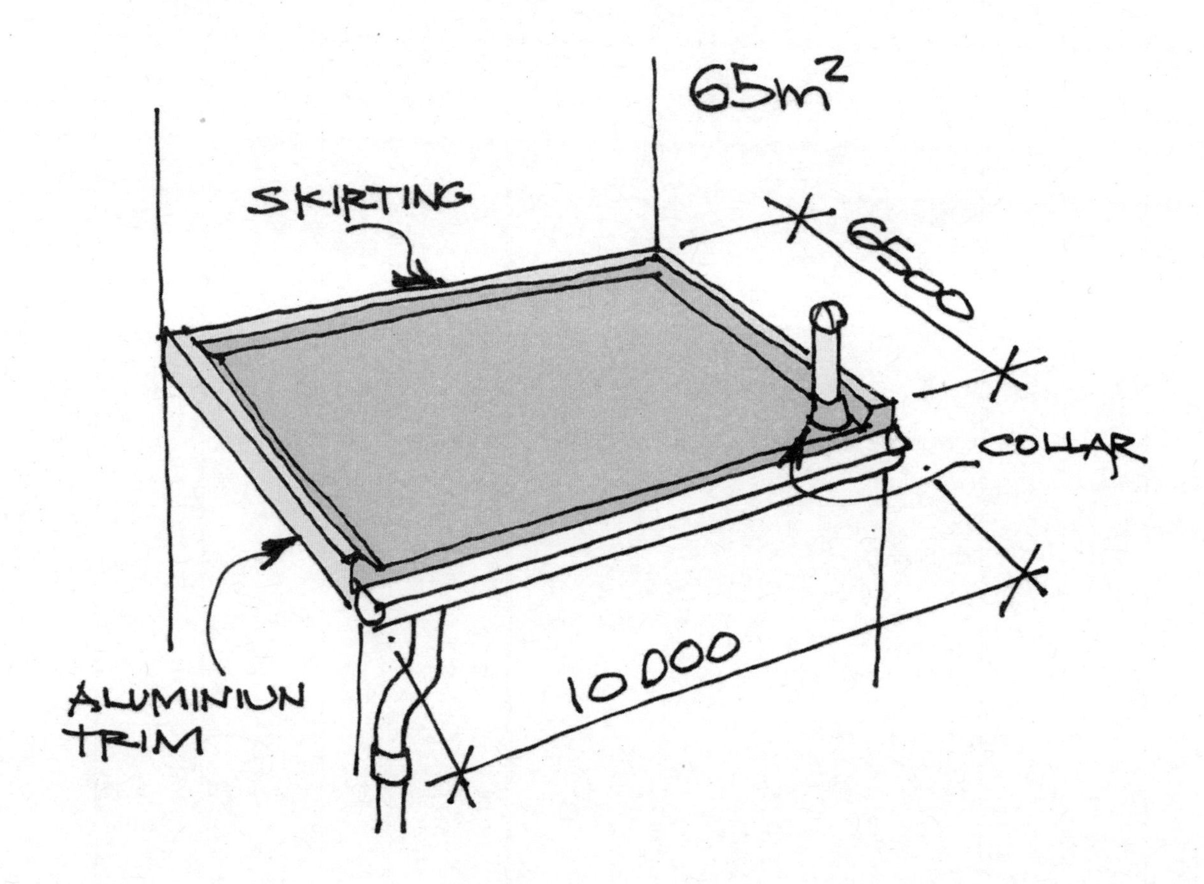

Signed for & on behalf of the PM

....

Date

....

Measured Term Contract Order No: *EXAMPLE BW15*

BUILDING 2009

Measurement & Valuation

				Item ref	*Unit*	*Quantity*	*Rate*	*Extension*
	10.30 6.80	70.04	Stripping felt roofing, to masonry, any number layers	C90.064	m²	70.04	3.13	219.23
	10.00 6.50	65.00	Three layer bitumen felt; roof covering horizontal or to falls, crossfalls or slopes ne 10° pitch — bottom layer	J41.001/4	m²	65.00	8.24	535.60
			intermediate layer	J41.001/4	m²	65.00	8.24	535.60
			cap sheet	J41.001/5	m²	65.00	14.29	928.85
			&					
			Insulation glassboard	J41.037/4	m²	65.00	18.39	1195.35
2 2	10.00 6.50	20.00 13.00 33.00	Turn-up or turn down at abutment, eaves or verges, skirting, flashing or apron; welted — bottom layer	J41.016/4	m	33.00	4.43	146.19
			intermediate layer	J41.016/4	m	33.00	4.43	146.19
			cap sheet	J41.016/5	m	33.00	5.36	176.88
	1	1	Collar; around pipe, standard or the like; ne 150 mm high; 55 to 110 mm dia — bottom layer	J41.032/4	Ea	1	6.82	6.82
			intermediate layer	J41.032/4	Ea	1	6.82	6.82
			cap sheet	J41.032/5	Ea	1	7.99	7.99
	10.00 6.50	65.00	ADD for priming base	J41.005/4	m²	65.00	2.15	139.75
			&					
			Chippings	J41.007/5	m²	65.00	5.78	375.70
			&					
			Cleaning base after removal of felt	J41.046	m²	65.00	0.98	63.70
	10.00	10.00	Turning back lead flashing & redress	H71.069	m	10.00	3.05	30.50
2	6.50	13.00	Aluminium eaves trim	J41.044/2	m	13.00	7.42	96.46
								4611.63
			Contract % adjustment			***-10.00%***		-461.16
								4150.47
			Add updating % June 2009			0.5%		20.75
			Net total					***£4,171.22***

Measured Term Contract Order No:- *EXAMPLE BW16*

Location:-

Work to be completed by:-

Please execute the following:-

TAKE OUT EXISTING SINK, TRAP AND STEEL WASTE PIPE. PROVIDE AND FIX IN POSITION NEW SINK WITH WASTE PIPE AND CONNECTION (ASSUMING NOMINAL HORIZ. DISTANCE 2m FOR WASTE PIPE)

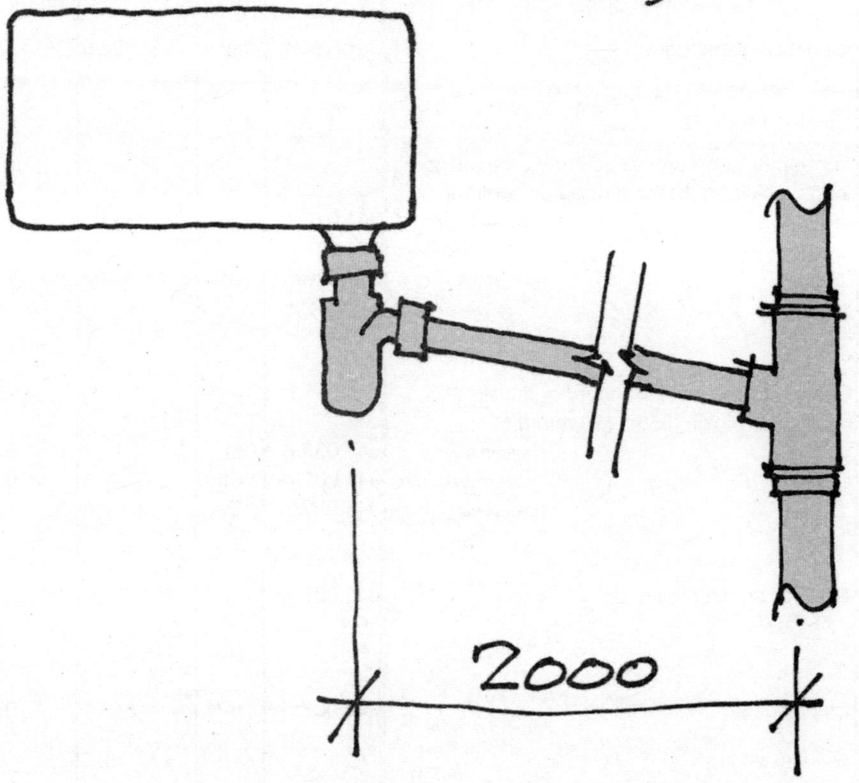

2000

Signed for & on behalf of the PM

....

Date

....

Measured Term Contract Order No: *EXAMPLE BW16*

BUILDING 2009

Measurement & Valuation

				Item ref	*Unit*	*Quantity*	*Rate*	*Extension*
	2	2	Cutting to release end of metal section ne 250mm depth, common bkwk	F10.254/1	Ea	2	9.22	18.44
	2.00	2.00	Take out steel waste pipe 25 to 50mm dia.	C90.215/2	m	2.00	4.84	9.68
	1	1	Take out trap	C90.168/1	Ea	1	10.66	10.66
	1	1	Burn out joint, 50mm steel pipe	C90.170	Ea	1	7.37	7.37
	1	1	Take out sink	N13.060/3	Ea	1	16.75	16.75
	2.00	2.00	42mm copper waste	R11.152/3	m	2.00	27.99	55.98
	2.00	2.00	ADD for brackets	R11.155/3	m	2.00	10.82	21.64
	1	1	Pipe connector	R11.162/3	Ea	1	40.04	40.04
	1	1	38mm copper P trap	R11.219/3	Ea	1	58.25	58.25
	1	1	Glazed fireclay sink 610 x 455 x 255mm	N13.060/1	Ea	1	230.50	230.50
	2	2	CI cantilever bracket	N13.068/1	Ea	2	12.25	24.50
								493.81
			Contract % adjustment			***-10.00%***		-49.38
								444.43
			Add updating % June 2009			0.5%		2.22
			Net total					***£446.65***

Measured Term Contract Order No:- *EXAMPLE BW17*

Location:-

Work to be completed by:-

Please execute the following:-

Take out basin, trap, steel waste and supply pipes. Provide and fix in position new basin with copper waste, hot and cold water supply pipes, bends, tees, fittings and connections (assume nominal horizontal distance of 2 m for supply and waste pipes)

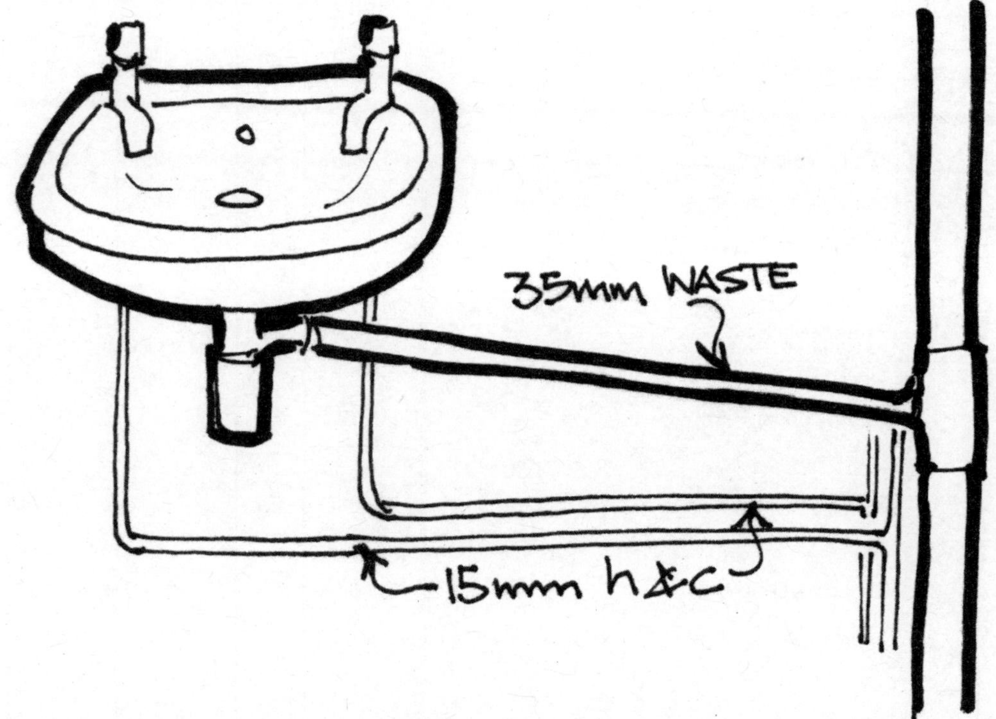

Signed for & on behalf of the PM

....

Date

....

Measured Term Contract Order No: *EXAMPLE BW17*

BUILDING 2009

Measurement & Valuation

				Item ref	Unit	Quantity	Rate	Extension
	6.50	6.50	Take out steel pipe 25 to 50mm dia.	C90.215/2	m	6.50	4.84	31.46
	1	1	Take out trap	C90.168/1	Ea	1	10.66	10.66
	1	1	Burn out joint, ne 50mm steel pipe	C90.170	Ea	1	7.37	7.37
	1	1	Take out wash basin	N13.043/3	Ea	1	14.31	14.31
	2.00	2.00	35mm copper waste	R11.152/2	m	2.00	22.88	45.76
	2.00	2.00	ADD for brackets	R11.155/2	m	2.00	9.52	19.04
	2	2	Made bend 35mm	R11.159/2	Ea	2	8.04	16.08
	1	1	Pipe connector 35mm	R11.162/2	Ea	1	33.97	33.97
	5.00	5.00	15mm copper supply	S10.107/1	m	5.00	6.96	34.80
	5.00	5.00	ADD for brackets	S10.111/1	m	5.00	6.84	34.20
	2	2	Made bend 15mm	S10.115/1	Ea	2	4.02	8.04
	2	2	Pipe connector 15mm	S10.116/1	Ea	2	8.32	16.64
	2	2	Fitting connector 15mm	S10.123/1	Ea	2	10.49	20.98
	1	1	38mm copper P trap	R11.219/3	Ea	1	58.25	58.25
	1	1	Wash basin, vit china 560 x 405mm	N13.043/1	Ea	1	97.25	97.25
	1	1	Basin support brackets (pair)	N13.049/1	Ea	1	23.38	23.38
	2	2	Pillar tap, 13mm (pair)	N13.009/1	Ea	1	41.67	41.67
								513.86
			Contract % adjustment			*-10.00%*		-51.39
								462.47
			Add updating % June 2009			0.5%		2.31
			Net total					***£464.78***

Measured Term Contract Order No:- *EXAMPLE BW18*

Location:-

Work to be completed by:-

Please execute the following:-

Take out WC pan, cistern, waste and supply pipes. Provide and fix in position new WC pan and cistern with cold water supply pipes, bends, tees, fittings and connections (assume nominal horizontal distance of 2 m for supply and waste pipes)

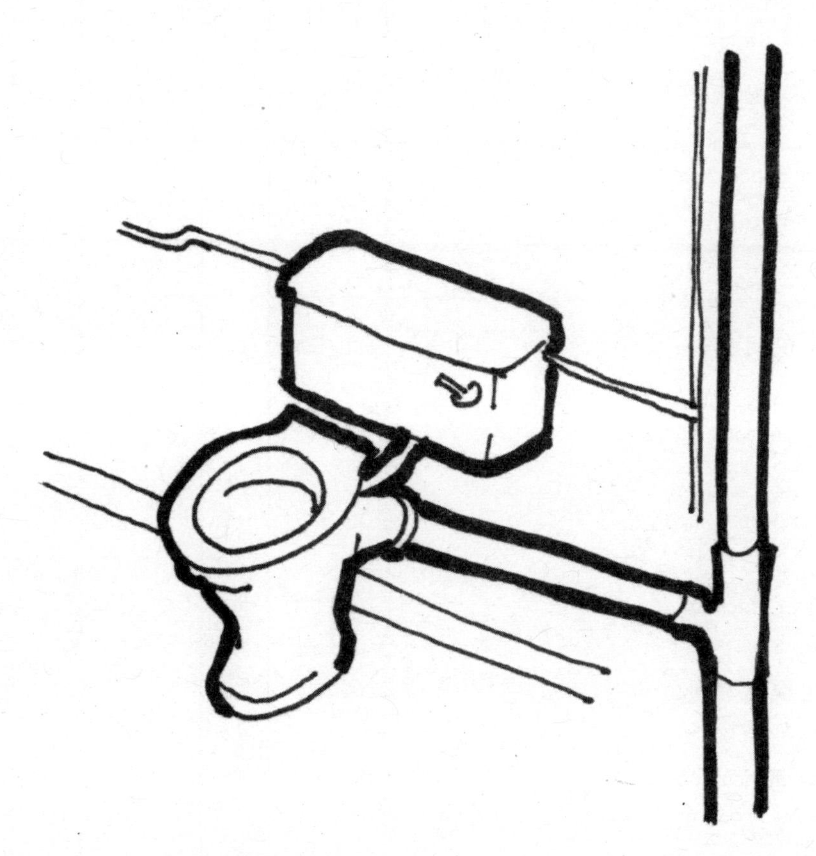

Signed for & on behalf of the PM

....

Date

....

Measured Term Contract Order No: *EXAMPLE BW18*

BUILDING 2009

Measurement & Valuation

				Item ref	Unit	Quantity	Rate	Extension
	1.22	1.22	Take out c.i. soil pipe 50 to 100 mm dia.	C90.157/2	m	1.22	3.90	4.76
	2.65	2.65	Take out steel pipe n.e. 25 mm	C90.215/1	m	2.65	3.05	8.08
	1	1	Burning out ex joint	C90.172	Ea	1	14.50	14.50
	1	1	Take out WC pan	N13.078/3	Ea	1	24.36	24.36
	1.22	1.22	CI spigot & socket soil pipe 100mm	R11.066/3	m	1.22	61.25	74.73
	1	1	Socket for stoneware 100mm	R11.073/3	Ea	1	58.14	58.14
	1	1	Bend 100mm	R11.085/3	Ea	1	63.26	63.26
	1	1	Connection to existing	R11.111/3	Ea	1	30.58	30.58
	2.00	2.00	22mm copper overflow	S10.107/2	m	2.00	10.03	20.06
	2.00	2.00	ADD for brackets	S10.111/2	m	2.00	7.14	14.28
	2	2	Made bend 22mm	S10.115/2	Ea	2	4.99	9.98
	1	1	Fitting connector 22mm	S10.123/2	Ea	1	12.31	12.31
	1	1	WC pan	N13.078/1	Ea	1	140.33	140.33
	1	1	Low level cistern	N13.088/1	Ea	1	81.70	81.70
	1	1	Black plastic seat	N13.094/1	Ea	1	24.05	24.05
	2.00	2.00	15mm copper supply	S10.107/1	m	2.00	6.96	13.92
	2.00	2.00	ADD for brackets	S10.111/1	m	2.00	6.84	13.68
	1	1	Pipe connector 15mm	S10.116/1	Ea	1	8.32	8.32
	1	1	Fitting connector 15mm	S10.123/1	Ea	1	10.49	10.49
								627.53
			Contract % adjustment			***-10.00%***		-62.75
								564.78
			Add updating % June 2009			0.5%		2.82
			Net total					***£567.60***

Measured Term Contract Order No:- *EXAMPLE BW19*

Location:-

Work to be completed by:-

Please execute the following:-

227 LITRE PLASTIC COLD WATER TANK ON 50 x 100 mm SW BEARERS AND 25mm SW BOARDING, POLYTHENE RISING MAIN, OVERFLOW AND SUPPLY PIPES, FITTINGS, COPPER BALL VALVE AND STOP VALVES, PREFORMED INSULATION TO TANK AND PIPES

SINGE STOREY BUILDING 3m HIGH

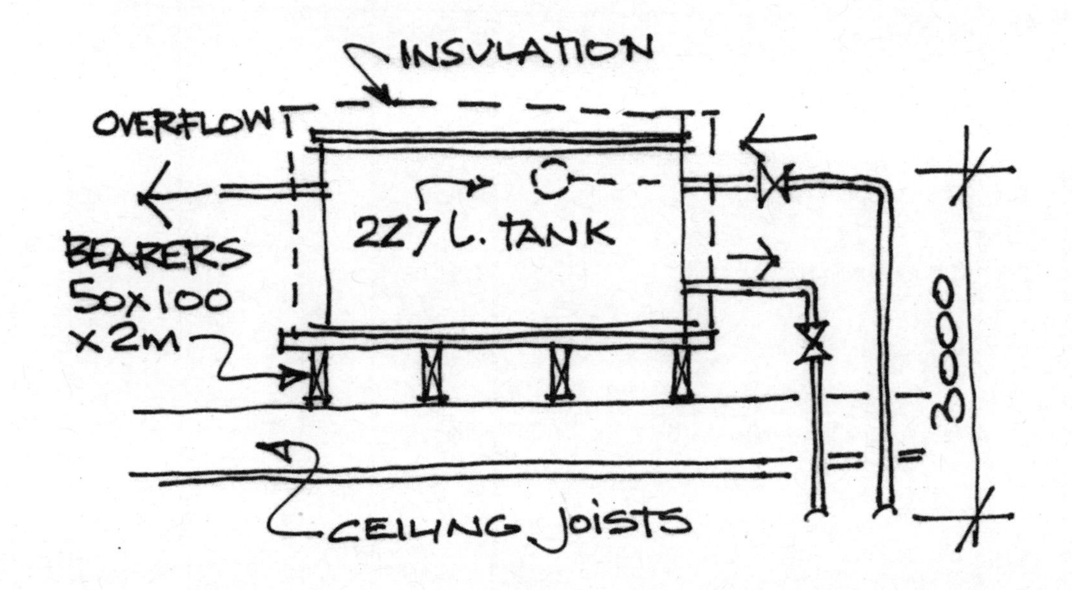

Signed for & on behalf of the PM

....

Date

....

Measured Term Contract Order No: *EXAMPLE BW19*

BUILDING 2009

Measurement & Valuation

				Item ref	Unit	Quantity	Rate	Extension
	1	1	Ball valve 19mm, plastics float	S10.196/8	Ea	1	34.78	34.78
4	2.00	8.00	Sawn softwood bearers 100 x 50 = 5000 m² £8.16 / 10000 x 5000 = £4.08	G20.005/1	m	8.00	4.08	32.64
	2.00 2.00	4.00	Sawn boarding 25 mm	K20.009/2	m²	4.00	16.63	66.52
	1	1	Pipe/tank connector 19mm	S10.032/1	Ea	1	14.10	14.10
	2	2	Pipe/tank connector 25mm	S10.032/2	Ea	2	19.42	38.84
	4.50	4.50	Polythene pipe, black 25 mm overflow	S10.023/2	m	4.50	7.58	34.11
	8.00	8.00	Polythene pipe, black 19 mm supply pipe	S10.023/1	m	8.00	5.47	43.76
	7.00	7.00	Polythene pipe, black 25 mm rising main	S10.023/2	m	7.00	7.58	53.06
	3.00	3.00	ADD for clips to masonry 19mm	S10.026/1	m	3.00	4.59	13.77
	2.00	2.00	ADD for clips to masonry 25mm	S10.026/2	m	2.00	4.71	9.42
	4	4	Made bend 19 mm	S10.029/1	Ea	4	8.68	34.72
	6	6	Made bend 25 mm	S10.029/2	Ea	6	12.79	76.74
	1	1	Stop valve, comp jnt to plastics 22 mm	S10.177/2	Ea	1	21.07	21.07
	2	2	Stop valve, comp jnt to plastics 28 mm	S10.177/3	Ea	2	33.54	67.08
	1	1	Plastics cistern 227 litre, rectangular	S10.241/1	Ea	1	132.54	132.54
	1	1	Cover to cistern	S10.241/2	Ea	1	36.64	36.64
	3	3	Perforation	S10.250	Ea	3	4.57	13.71

continued over page

continued from previous page

2	1.30		Insulation to cistern 40 mm thick	S10.302/1	m^2	2.54	15.19	38.51
	0.65	1.69						
2	0.65							
	0.65	0.85						
		2.54						
	3	3	Fitting around pipe	S10.303/1	Ea	3	1.52	4.56
	10.00	10.00	Insulation to pipe, 25mm th	S10.292/5	m	10.00	15.18	151.80
	2	2	Working around ancillary	S10.298/5	Ea	2	10.81	21.62
								939.99
			Contract % adjustment			***-10.00%***		-94.00
								845.99
			Add updating % June 2009			0.5%		4.23
			Net total					***£850.22***

Measured Term Contract Order No:- *EXAMPLE BW20*

Location:-

Work to be completed by:-

Please execute the following:-

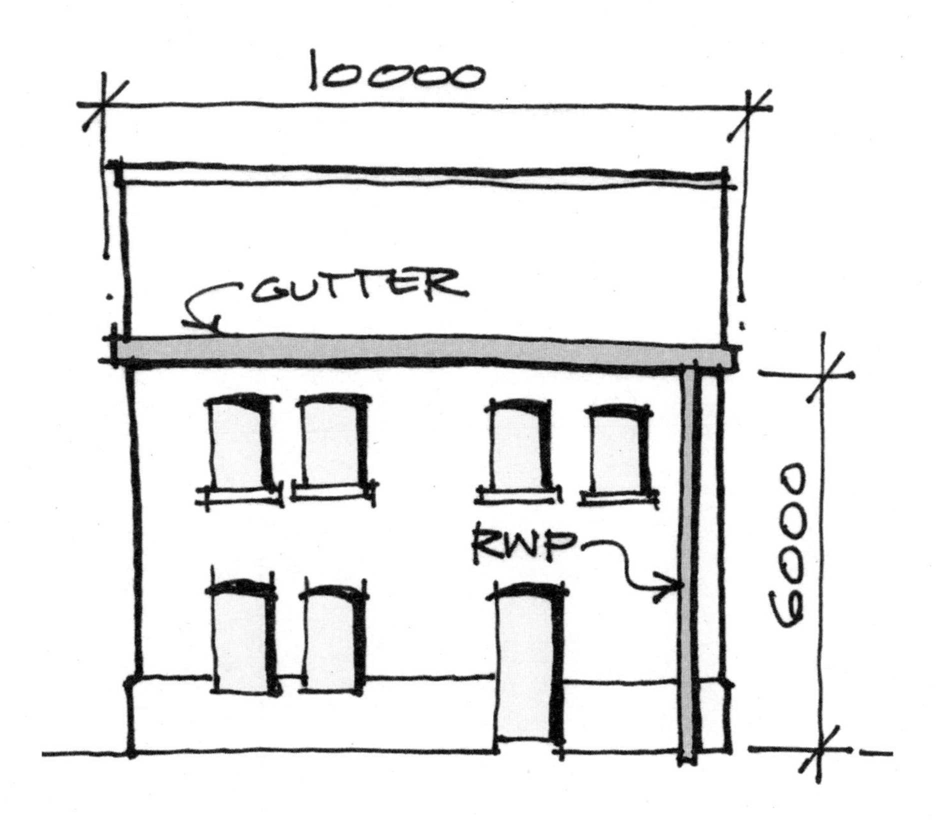

Signed for & on behalf of the PM

....

Date

....

Measured Term Contract Order No: *EXAMPLE BW20*

BUILDING 2009

Measurement & Valuation

			Description	*Item ref*	*Unit*	*Quantity*	*Rate*	*Extension*
	6.00	6.00	Take down existing rainwater pipe 50 to 100 mm dia	C90.151/2	m	6.00	3.01	18.06
			&					
			PVC rainwater pipe 68mm dia	R10.020/2	m	6.00	7.26	43.56
			&					
			Brackets 68 mm dia to masonry	R10.022/2	m	6.00	5.63	33.78
	1	1	Offset bend, 300mm projection 68 mm dia	R10.033/2	Ea	1	7.09	7.09
	1	1	Jointing to extg 68 mm dia	R10.038/2	Ea	1	3.29	3.29
	1	1	Wire balloon	R10.119/2	Ea	1	3.25	3.25
	10.00	10.00	Take down existing gutter ne 100mm w	C90.159/1	m	10.00	3.62	36.20
			&					
			PVC 100 mm half round gutter	R10.134/1	m	10.00	14.36	143.60
	2	2	Stopped end	R10.135/1	Ea	2	4.41	8.82
	1	1	Outlet	R10.137/1	Ea	1	10.50	10.50
								308.15
			Contract % adjustment			***-10.00%***		-30.82
								277.33
			Add updating % June 2009			0.5%		1.39
			Net total					***£278.72***

Measured Term Contract Order No:- *EXAMPLE BW21*

Location:-

Work to be completed by:-

Please execute the following:-

REPLACE DEFECTIVE 150mm CLAYWARE PIPE BETWEEN S.V.P AND MANHOLE (ASSUME 2m)
AVERAGE DEPTH 750mm.

2000

NB. ALL DIMS ARE IN mm.

Signed for & on behalf of the PM

....

Date

....

Measured Term Contract Order No: *EXAMPLE BW21*

BUILDING 2009

Measurement & Valuation

				Item ref	Unit	Quantity	Rate	Extension
	2.00	2.00	T u 150 mm clay pipe	C90.179/2	m	2.00	2.05	4.10
			&					
			Exc tr av 750 mm deep	R12.003/1	m	2.00	16.44	32.88
			&					
			Granular bed	R12.020/1	m	2.00	5.72	11.44
			&					
			Vit clay plain ended pipe 150 mm	R12.050/2	m	2.00	19.60	39.20
	2	2	Connect new pipe to 150 mm existing	R12.119/2	Ea	2	40.31	80.62
								168.24
			Contract % adjustment			***-10.00%***		-16.82
								151.42
			Add updating % June 2009			0.5%		0.76
			Net total					***£152.18***

Measured Term Contract Order No:- *EXAMPLE BW22*

Location:-

Work to be completed by:-

Please execute the following:-

NEW MANHOLE, INTERNAL SIZE 800x 675mm, 1000mm DEEP, NEW 100mm CLAYWARE DRAIN.

150mm PLAIN CONC. BED
250mm R.C. SLAB
ONE BRICK THICK WALL IN ENG. BRICKS (PC £360 PER 1000)
600 x 450mm CAST IRON MANHOLE COVER

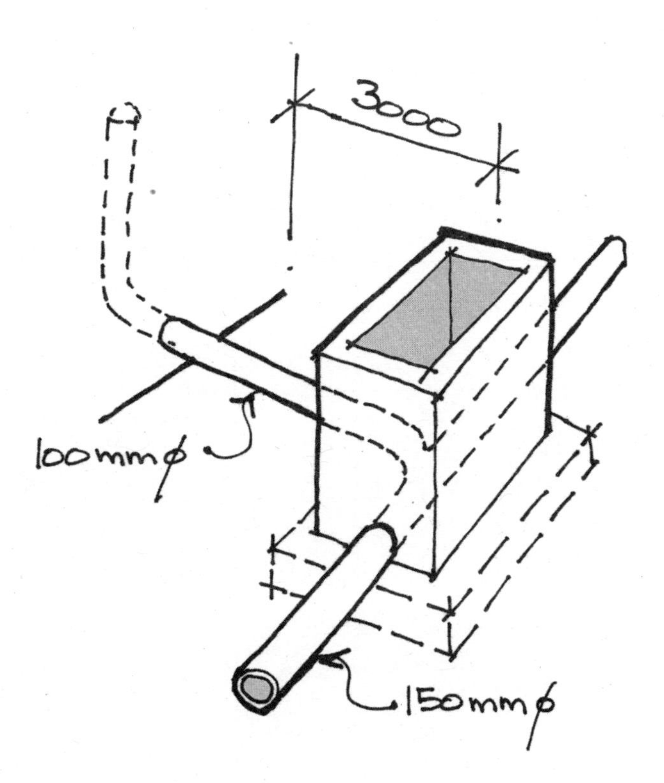

NB: ALL DIMS ARE IN mm

Signed for & on behalf of the PM

....

Date

....

Measured Term Contract Order No: *EXAMPLE BW22*

BUILDING 2005

Measurement & Valuation

					Item ref	Unit	Quantity	Rate	Extension
	1.25		Exc pit/ stg gl/ ne 2m dp	0.800+0.225*2					
	1.13			0.675+0.225*2					
	1.10	1.55		1.00+0.100	R12.250/2	m³	1.55	62.58	97.00
			Disp exc mat/ from site		R12.250/2	m³	1.55		Included
	1.25		Levelling etc bttm of exc						
	1.13	1.41			R12.251	m²	1.41	1.23	1.73
2	1.25		Earthwk sppt/ opp faces ne 2m /						
	1.10	2.75	max dpth ne 2m						
2	1.13								
	1.10	2.49							
		5.24			R12.252/2	m²	5.24	6.40	33.54
	1.25		Conc bed ne 150 mm th						
	1.13								
	0.15	0.21			R12.261	m³	0.21	142.09	30.11
	0.90		Benching in bttm mh/ av th 225 mm						
	0.57	0.51			R12.258	m²	0.51	69.58	35.49
	1.25		Slab/ reinf/ 150-450mm th/ 30N/ 20mm agg/						
	1.13		ord cem						
	0.25	0.35			R12.264	m³	0.35	160.53	56.19
4	1.20	4.80	16 mm h.y. bar reinforcement						
5	1.08	5.40							
		10.20	10.20	1.579	E30.005/1	m	16.11	1.06	17.07
11	1.20	13.20	12 mm m.s. bar reinforcement						
12	1.08	12.96							
		26.16	26.16	0.888	E30.013/1		23.23	1.31	30.43
	0.80		Sof slab/ 200mm th/ basic fin						
	0.68	0.54			R12.266	m²	0.54	45.83	24.75
			ADD for additional 50mm thickness £3.83 x 50/100mm = £1.915		R12.267	m²	0.54	1.915	1.03
2	1.25	2.50	Edge susp slab/ ne 250mm hi/ basic fin						
2	1.13	2.26							
		4.76			R12.268	m	4.76	15.51	73.83

continued over page

continued from previous page

2	1.03		1B wall/ Eng. bks/ mh					
	0.75	1.55						
2	0.90							
	0.75	1.35						
		2.90		R12.301/2	m^2	2.90	90.05	261.15
2	0.80		Extra for facewk					
	0.53	0.84						
2	0.68							
	0.53	0.71						
		1.55		R12.305/2	m^2	1.55	9.85	15.27
	1	1	150 mm h.r. channel 900 mm long	R12.315/2	Ea	1	15.43	15.43
			0.90 - 0.60 = 0.30 / 0.15 = 2	R12.316/2	Ea	2	4.06	8.12
	2	2	100 mm h.r. channel branch bend	R12.322/1	Ea	2	21.00	42.00
	1	1	Cut out old 150 mm clay pipe	R12.378/3	Ea	1	118.10	118.10
	1	1	Bldg in end of 100mm pipe/ Eng bks/ m g	R12.308/2	Ea		7.00	
				R12.310	factor		2.00	
						1	14.00	14.00
	2	2	Bldg in end of 150mm pipe/ Eng bks/ m g	R12.308/3	Ea		10.50	
				R12.310	factor		2.00	
						2	21.00	42.00
	1	1	600 x 450 mm c.i. m.h. cover/ set in c.m.	R12.337	Ea	1	150.66	150.66
	1.60	1.60	Exc tr av 1000 mm deep	R12.004/1	m	1.60	22.09	35.34
			&					
			Granular bed	R12.020/1	m	1.60	5.72	9.15
			&					
			Vit clay plain ended pipe 100 mm	R12.050/1	m	1.60	11.53	18.45
								1130.83
			Contract % adjustment			***-10.00%***		-113.08
								1017.74
			Add updating % June 2009			0.5%		5.09
								1022.83
			ADD bricks (as per invoice)					
			2.90m² x 128 = 372		Thou	***0.372***	***360.00***	133.92
					Profit		***5.00%***	6.70
			Net total					***£1,163.45***

Measured Term Contract Order No:- *EXAMPLE BW23*

Location:-

Work to be completed by:-

Please execute the following:- Strip foundation as detailed below

Deposit surplus excavated material on site - 100m from site of excavation

Normal water table occurs at 0.6m below ground level

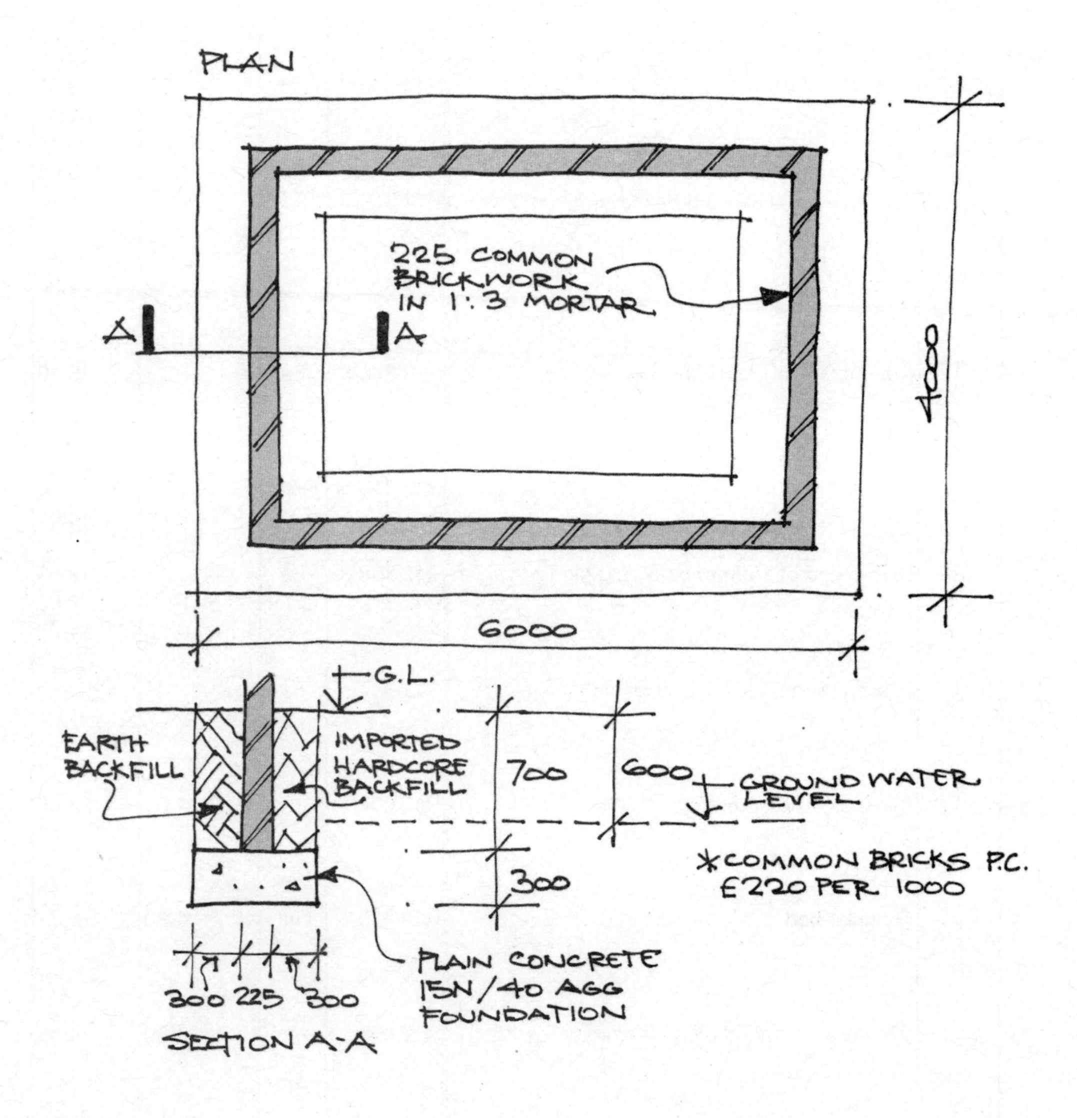

Signed for & on behalf of the PM

....

Date

....

Measured Term Contract Order No:- *EXAMPLE 23*

BUILDING 2009

THIS VALUATION CONTAINS DELIBERATE ERRORS FOR TRAINING PURPOSES

Measurement & Valuation

				Item ref	Unit	Quantity	Rate	Extension
			Excavation					
			6.000 4.000 10.000 x 2 = 20.000					
	20.00 0.83 1.00	16.60	Exc trch/ stg gl/ ne 2m dp/ over 0.30m wide &	D20.013/3	m³	17.00	19.00	323.00
			Disp exc mat/ on site/ 100m dist D44 x 100/50 x 0.85 = 1.70	D20.043 D20.044	m³ m³	17.00	8.33 1.70 10.03	170.51
	20.00 0.83 0.60	9.96	EO for excg bel grd water lev	D20.018	m³	10.00	4.09	40.90
3	8	24	Pumping - single diaphram 100ø	D20.058	Hrs	24.00	7.48	179.52
	20.00 0.83	16.60	Levelling etc bttm of exc	D20.053	m²	17.00	0.59	10.03
	20.00 0.30 0.70	4.20	Backfilling: earth from excavations &	D20.046/2	m³	5.00	14.34	71.70
			imported hardcore	D20.046/6	m³	5.00	45.69	228.45
2.00	20.00 1.00	40.00	Earthwork support Opp faces ne 2m/ max dpth ne 2m extending below ground water level	D20.032/2 D20.036/2	m² m²	40.00	6.40 11.19 17.59	703.60
	20.00 0.83 0.30	4.98	Concrete Concrete in foundations	E10.005/1	m³	5.00	134.34	671.70
	20.00 0.70	14.00	Brickwork Laying only - one brick thick in cm 1:3	F10.063/2	m²	14.00	92.29	1292.06
								3691.47
			Contract % adjustment			-10.00%		-369.15
								3322.32
			ADD bricks (as per invoice) 14m² x 2 x 65 = 1820 Plus waste say 2000		Thou Profit	2.00	220.00 10.00%	440.00 44.00
			Daywork - Attendance on pumping B Rickie (bricklayer)		hrs	6.00	10.58	63.48
								3806.32
			Add updating % June 2009			0.5%		19.03
			Net total					**£3,825.35**

Measured Term Contract Order No:- *EXAMPLE 23*

BUILDING 2009

Measurement & Valuation

			Description	*Item ref*	*Unit*	*Quantity*	*Rate*	*Extension*
			Excavation					
			6.000					
			4.000				£	£
			10.000 x 2 = 20.000					
			Less 4/2/0.5/0.825 = 3.300					
			= ***16.700***					
	16.70		Exc trch/ stg gl/ ne 1m dp/ over 0.30m wide	***D20.014/2***	m³	***13.86***	***13.52***	187.40
	0.83							
	1.00	***13.86***	&					
			Disp exc mat/ on site/ 100m dist	D20.043	m³		8.33	
			D44 x 75/50 x 0.85 = 1.28	D20.044	m³		***1.28***	
			13.86m³ less backfill 3.95m³ = 9.91m³			***9.91***	***9.61***	95.19
	16.70		EO for excg bel grd water lev	D20.018	m³	***5.54***	4.09	22.68
	0.83		***Pumping - not allowable***					
	0.40	5.54						
	16.70		Levelling etc bttm of exc					
	0.83	***13.86***		D20.053	m²	***13.86***	0.59	8.18
			6.000					
			4.000					
			10.000 x 2 = 20.000					
			Less 4/2/0.5/0.300 = 1.200					
			= ***18.800***					
	18.80		Backfilling: earth from excavations	D20.046/2	m³	***3.95***	14.34	56.61
	0.30							
	0.70	***3.95***	&					
			6000-2/0.825 4.350					
			4.000-2/0.825 2.350					
			6.700 x 2 = 13.400					
			Add 4/2/0.5/.300 = 1.200					
			= ***14.600***					
	14.60		Backfilling: imported hardcore	D20.046/6	m³	***3.07***	45.69	140.09
	0.30							
	0.70	***3.07***						
			Earthwork support					
2.00	***16.70***		Opp faces ne 2m/ max dpth ne 1m	D20.032/1	m²		***5.53***	
	1.00	***33.40***	extending below ground water level	D20.036/1	m²		***10.12***	
						33.40	***15.65***	522.71
			Concrete					
	16.70		Concrete in foundations	E10.005/1	m³		134.34	
	0.83		***Adjust for mix***	***E10.029/1***			***-4.38***	
	0.30	***4.16***				***4.16***	***129.96***	540.41
			Brickwork					
	16.70		Laying only - one brick thick in cm 1:3	***F10.001/2***	m²		***72.47***	
	0.70	***11.69***		***F10.015***			***3.00***	
						11.69	***75.47***	882.24
								2455.51
			Contract % adjustment			***-10.00%***		-245.55
								2209.96
			Add updating % June 2009			0.5%		11.05
								2221.01
			ADD bricks (as per invoice)					
			11.69m² x 128 = 1497		Thou	***1.497***	***220.00***	329.34
					Profit		***5.00%***	16.47
			Net total					***£2,566.81***

Measured Term Contract Order No:- *EXAMPLE BW24*

Location:-

Work to be completed by:-

Please execute the following:-

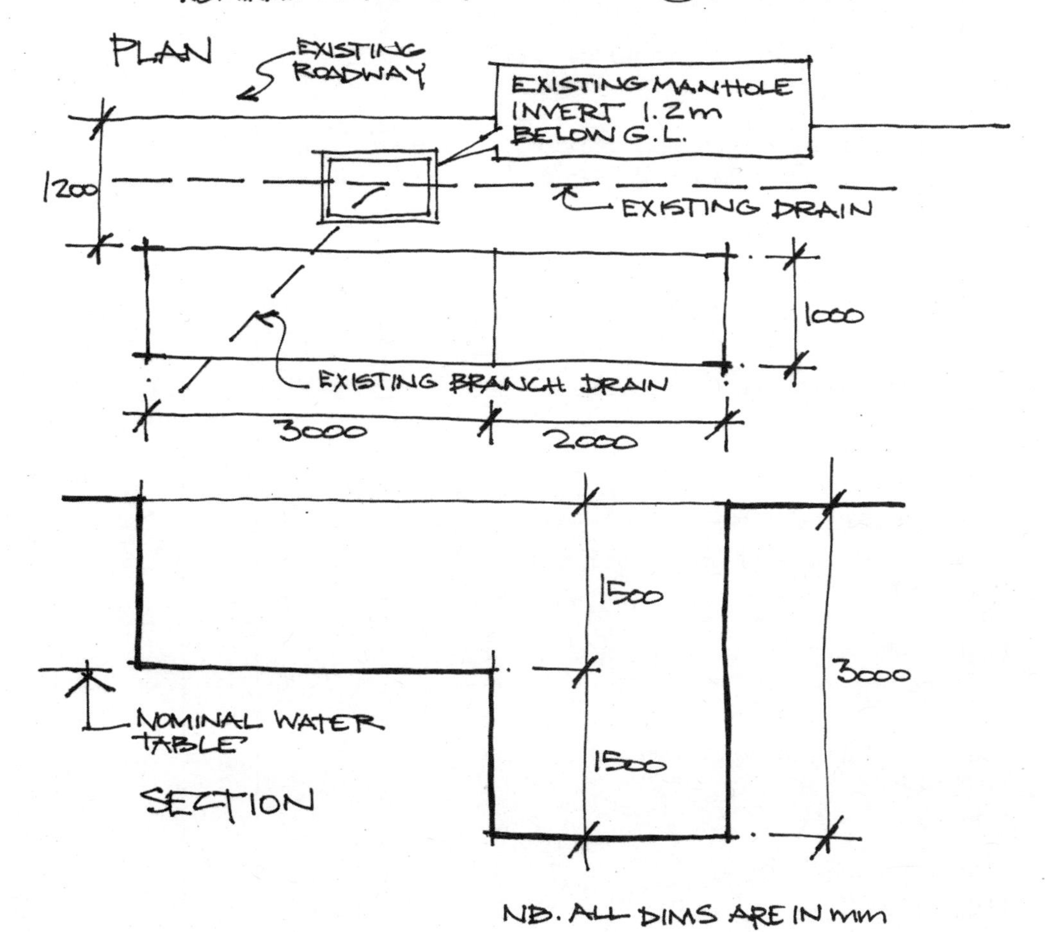

Signed for & on behalf of the PM

....

Date

....

Measured Term Contract Order No: *EXAMPLE 24*

BUILDING 2009

Measurement & Valuation

				Item ref	*Unit*	*Quantity*	*Rate*	*Extension*
			Excavation					
	5.00		Exc trch/ stg gl/ ne 2m dp/ over 0.30m wide					
	1.00							
	1.50	7.50		D20.014/3	m³	7.50	14.46	108.45
	2.00		Exc pit/ stg 1.50m bel gl/ ne 2m dp	D20.012/3	m³		15.85	
	1.00		D17 x 1.50 = £6.98	D20.017	m³		6.98	
	1.50	3.00				3.00	22.83	68.48
			&					
			EO for excg bel grd water lev	D20.018	m³	3.00	4.09	12.27
	as	7.50	Disp exc mat/ on site/ 350m dist	D20.043	m³		8.33	
	above	3.00	D44 x 325/50 = £5.53	D20.044	m³		5.53	
		10.50				10.50	13.86	145.48
	5.00		Levelling etc bttm of exc					
	1.00	5.00		D20.053	m²	5.00	0.59	2.95
	5.00	5.00	Ex for excg next service	D20.030	m	5.00	12.87	64.35
	1.00	1.00	Ex for excg across service	D20.031	Ea	1.00	38.53	38.53
			Earthwork support					
2.00	3.00		Opp faces ne 2m/ max dpth ne 2m					
	1.50	9.00						
	1.00							
	1.50	1.50						
		10.50		D20.032/2	m²	10.50	6.40	67.20
2.00	2.00		Opp faces ne 2m/ max dpth ne 4m/ bel gwl					
	3.00	12.00						
	1.00							
	3.00	3.00						
	1.00			D20.032/3	m²		7.51	
	1.50	1.50		D20.036/3	m²		12.47	
		16.50				16.50	19.98	329.67
	3.00		Add for support next roadway/					
	1.50	4.50	max depth ne 2m	D20.038/2	m²	4.50	7.40	33.30
	2.00		Add for support next roadway/					
	3.00	6.00	max depth ne 4m	D20.038/3	m²	6.00	7.52	45.12
								915.79
			Contract % adjustment			***-10.00%***		-91.58
								824.21
			Add updating % June 2009			0.5%		4.12
			Net total					***£828.33***

Measured Term Contract Order No:- *EXAMPLE BW25*

Location:-

Work to be completed by:-

Please execute the following:- Concrete Work only

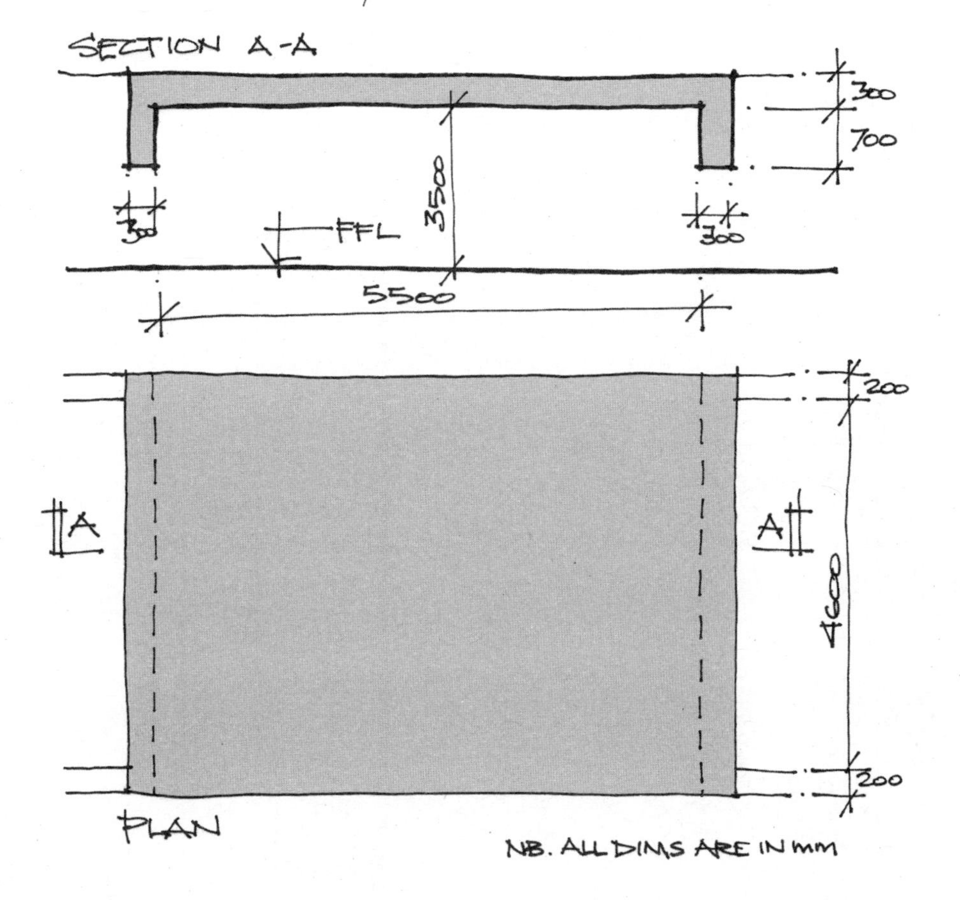

Signed for & on behalf of the PM

....

Date

....

Measured Term Contract Order No: *EXAMPLE 25*

BUILDING 2009

Measurement & Valuation

				Item ref	*Unit*	*Quantity*	*Rate*	*Extension*
			Concrete					
	6.00		Slab/ reinf/ 150-450mm th/ 30N/ 10mm agg/					
	5.00		ord cem					
	0.30	9.00						
	5.00							
	0.30			E10.002/5	m³		159.46	
	0.70	1.05		E10.037/1	m³		10.50	
		10.05				10.05	169.96	1708.10
	5.00		Attchd dp bm/ reinf/ 150-450mm th/ 30N/	E10.010/2	m³		185.32	
	0.20		10mm agg/ ord cem	E10.037/1	m³		10.50	
	0.70	0.70				0.70	195.82	137.07
			Formwork					
	5.50		Sof slab/ 300mm th/ 3.0-4.5m hi/ pl fin	E20.008	m²		45.46	
	4.60	25.30		E20.010	m²		3.83	
							49.29	
				E20.045	factor		1.125	
							55.45	
				E20.041	m²		4.90	
						25.30	60.35	1526.89
2	6.00	12.00	Edge susp slab/ 250-500mm hi/ pl fin	E20.003/3	m		25.24	
				E20.045	factor		1.125	
						12.00	28.40	340.74
2	4.60		Attchd beam/ pl fin	E20.016	m²		62.15	
	1.00	9.20		E20.045	factor		1.125	
2	4.60					17.94	69.92	1254.34
	0.70	6.44						
	4.60							
	0.30	1.38						
	4.60							
	0.20	0.92						
		17.94						
			FO Reinforcement					
	As		Y12/ str	E30.004/3	kg	500.00	0.66	330.00
	per		Y16/ str	E30.005/3	kg	700.00	0.58	406.00
	order		Y12/ bent	E30.004/4	kg	300.00	0.82	246.00
			R8/ links	E30.018/2	kg	100.00	1.15	115.00
			Sundries					
	6.00		Float fin	E41.002	m²	30.00	3.07	92.10
	5.00	30.00						
								6156.24
			Contract % adjustment			*-10.00%*		-615.62
								5540.62
			Add updating % June 2009			0.5%		27.70
			Net total					***£5,568.32***

Measured Term Contract Order No:- *EXAMPLE E1*

Location:-

Work to be completed by:-

Please execute the following:-

Replace fluorescent lamps in twin 36W 1200 mm batten luminaire opal diffuser with white triphosphor lamps.

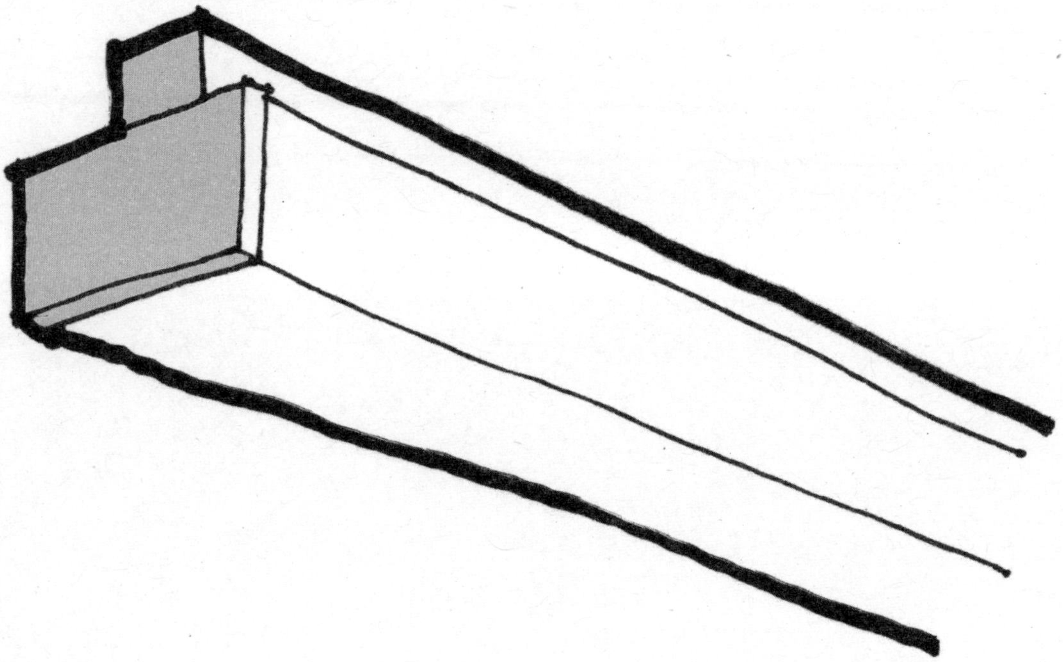

Signed for & on behalf of the PM

....

Date

....

Measured Term Contract Order No:-

ELECTRICAL 2006 - EXAMPLE 1

Measurement & Valuation

		Description	*Item ref*	*Unit*	*Quantity*	*Rate* A	*Rate* BDE	*Rate* C	*Extension* A	*Extension* BDE	*Extension* C
1	1	T/D extg twin opal diffuser to 36W 1200 mm batten fluorescent luminaire: for re-use	Y73.010/7	Ea factor	1		2.35 0.70 1.65			1.65	
2	2	T/D extg fluorescent lamps for disposal	Y73.967/1	Ea factor	2		2.35 0.40 0.94			1.88	
2	2	Supply 36W 1200 mm triphosphor white fluorescent lamp	Y73.942/1	Ea	2	3.59			7.18		
2	2	Fix ditto	Y73.967/1	Ea	2		2.35			4.70	
1	1	Re-fix extg twin opal diffuser to 36W 1200 mm batten fluorescent luminaire:	Y73.010/7	Ea	1		2.35			2.35	
									7.18	10.58	0.00
			Updating June 2009			A 15.5%	BDE 18.9%	C 15.9%	1.11	2.00	0.00
									8.29	12.58	0.00
			Contract % adjustment			A -10.0%	BDE -10.0%	C -10.0%	-0.83	-1.26	0.00
									7.46	11.32	0.00
									Grand total		**18.78**

Measured Term Contract Order No:- *EXAMPLE E2*

Location:-

Work to be completed by:-

Please execute the following:-

Renew batten high frequency luminaire twin 1800 mm 70W.

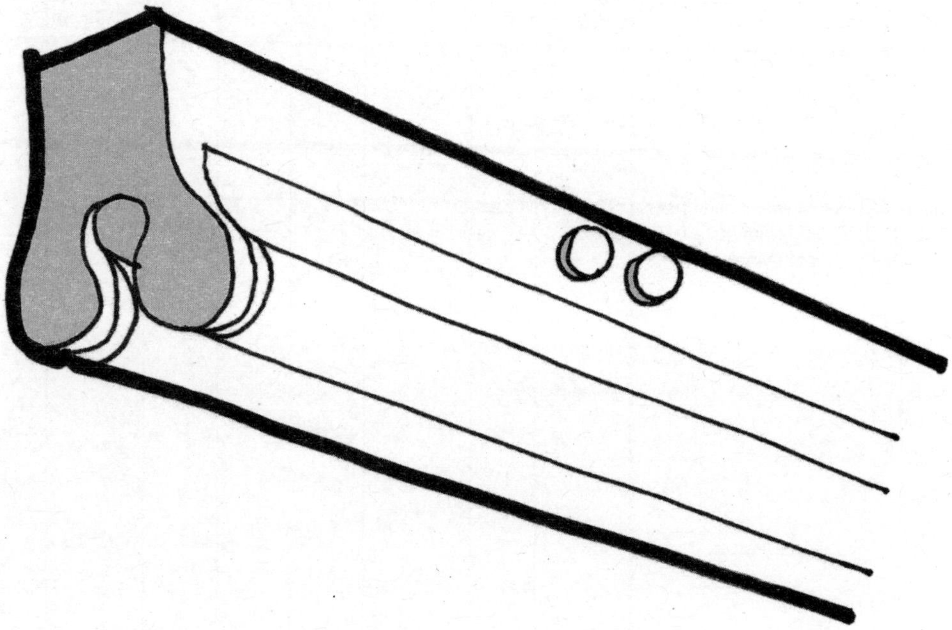

Signed for & on behalf of the PM

....

Date

....

Measured Term Contract Order No:-

ELECTRICAL 2006 - EXAMPLE 2

Measurement & Valuation

							Rate			*Extension*		
				Item ref	*Unit*	*Quantity*	*A*	*BDE*	*C*	*A*	*BDE*	*C*
			Batten fluorescent luminaire high frequency twin 1800mm 70W									
	1	1	Dismantle & remove	Y73.006/9	Ea			21.19				
					factor			0.40				
						1		8.48			8.48	
	1	1	Supply	Y73.006/4	Ea	1	54.72			54.72		
			Fix	Y73.006/9	Ea	1		21.19			21.19	
	2	2	Supply 70W 1800mm fluorescent lamp	Y73.935/1	Ea	2	1.26			2.52		
	2	2	Instal 70W 1800mm fluorescent lamp	Y73.967	Ea	2		2.35			4.70	
										57.24	34.37	0.00
				Updating June 2009						8.87	6.50	0.00
							A	BDE	C			
							15.5%	18.9%	15.9%			
										66.11	40.87	0.00
				Contract % adjustment						-6.61	-4.09	0.00
							A	BDE	C			
							-10.0%	-10.0%	-10.0%			
										59.50	36.78	0.00
										Grand total		**96.28**

Measured Term Contract Order No:- *EXAMPLE E3*

Location:-

Work to be completed by:-

Please execute the following:-

Take down 2 No. 3 x 18W 600 x 600 mm high frequency recessed modular fluorescent luminaire body in ceiling grid.

Re-fix 1 no.
Supply and fix 1 no. new

Supply and fix new general purpose lamps and Category 2 louvres

Signed for & on behalf of the PM

....

Date

....

Measured Term Contract Order No:-

ELECTRICAL 2006 - EXAMPLE 3

Measurement & Valuation

							Rate			*Extension*		
				Item ref	*Unit*	*Quantity*	*A*	*BDE*	*C*	*A*	*BDE*	*C*
	1	1	T/D extg 3 x 18W 600 x 600 recessed modular fluorescent luminaire: for disposal	Y73.030/5	Ea factor	1		28.72 0.40 11.49			11.49	
	1	1	T/D extg 3 x 18W 600 x 600 recessed modular fluorescent luminaire for re-use	Y73.030/5	Ea factor	1		28.72 0.70 20.10			20.10	
	1	1	Supply 3 x 18W 600 x 600 recessed modular fluorescent luminaire	Y73.030/1	Ea	1	43.79			43.79		
	2	2	Instal 3 x 18W 600 x 600 recessed modular fluorescent luminaire	Y73.030/5	Ea	2		28.72			57.44	
2	3	6	Supply 18W 600mm fluorescent lamp	Y73.928/1		6	0.69			4.14		
2	3	6	Instal 18W 600mm fluorescent lamp	Y73.967		6		2.35			14.10	
	2	2	Supply Cat 2 louvre	Y73.034/1	Ea	2	24.57			49.14		
	2	2	Instal Cat 2 louvre	Y73.034/5	Ea	2		7.30			14.60	
										97.07	117.73	0.00
				Updating June 2009			A 15.5%	BDE 18.9%	C 15.9%	15.05	22.25	0.00
										112.12	139.98	0.00
				Contract % adjustment			A -10.0%	BDE -10.0%	C -10.0%	-11.21	-14.00	0.00
										100.91	125.98	0.00
										Grand total		**226.89**

Measured Term Contract Order No:- *EXAMPLE E4*

Location:-

Work to be completed by:-

Please execute the following:-

Replace defective standard bathroom low voltage centrifugal fan 100 mm dia. with new including pull cord switch.

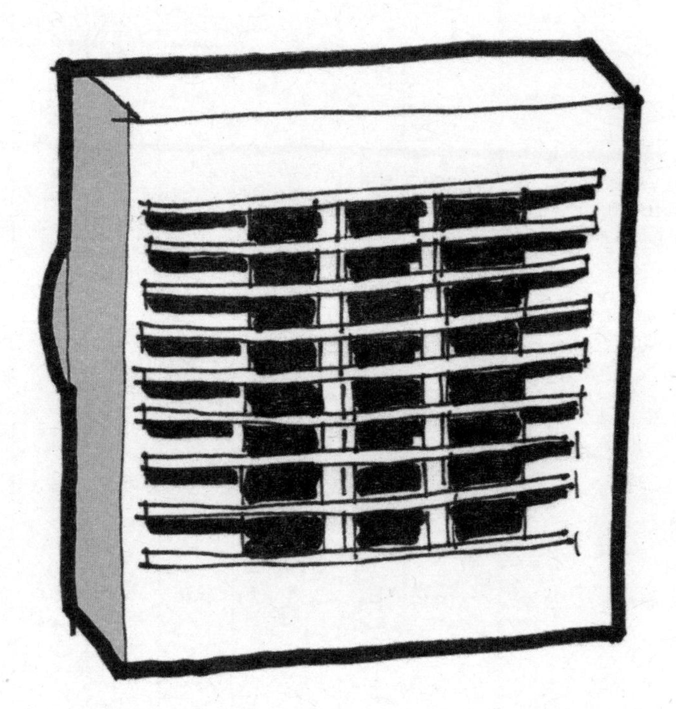

Signed for & on behalf of the PM

....

Date

....

Measured Term Contract Order No:-

ELECTRICAL 2006 - EXAMPLE 4

Measurement & Valuation

							Rate			Extension		
				Item ref	*Unit*	*Quantity*	*A*	*BDE*	*C*	*A*	*BDE*	*C*
	1	1	Dismantle and remove bathroom fan and remove for scrap	Y41.008/4	Ea factor	1		42.82 0.75 32.12			32.12	
	1	1	Bathroom fan, centrifugal, pull cord, low voltage supply	Y41.008/2	Ea	1	82.19			82.19		
	1	1	fix	Y41.008/4	Ea	1		42.82			42.82	
										82.19	74.94	0.00
				Updating June 2009			A 15.5%	BDE 18.9%	C 15.9%	12.74	14.16	0.00
										94.93	89.10	0.00
				Contract % adjustment			A -10.0%	BDE -10.0%	C -10.0%	-9.49	-8.91	0.00
										85.44	80.19	0.00
										Grand total		**165.63**

Measured Term Contract Order No:- *EXAMPLE E5*

Location:-

Work to be completed by:-

Please execute the following:-

STRIP OUT EXISTING AND REPLACE WITH NEW PVC INSULATED 2·5mm² 2-CORE PVC SHEATHED CABLE 350/500V FIXED BETWEEN FITTING AND EQUIPMENT (10 METRES)

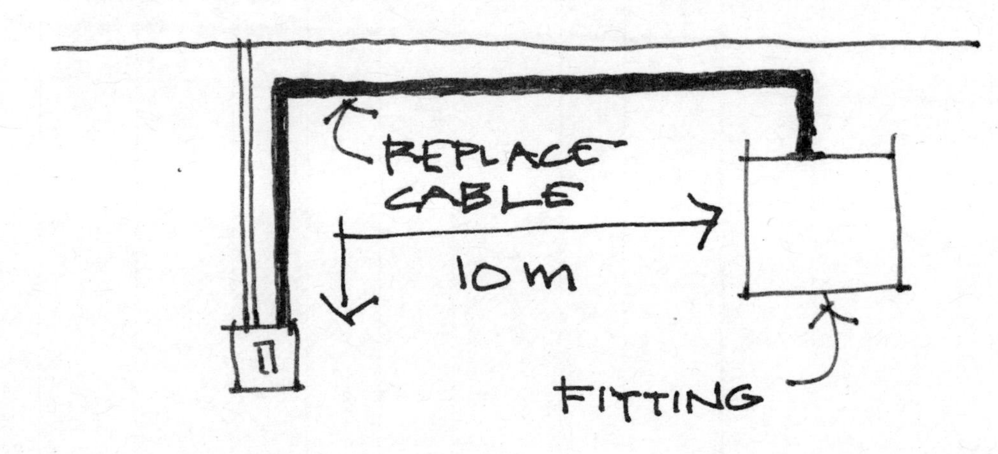

Signed for & on behalf of the PM

....

Date

....

Measured Term Contract Order No:-

ELECTRICAL 2006 - EXAMPLE 5

Measurement & Valuation

							Rate			*Extension*		
				Item ref	*Unit*	*Quantity*	*A*	*BDE*	*C*	*A*	*BDE*	*C*
	10.00	10.00	Strip out extg cable	Y61.095/3	m			3.18				
	0.30	0.30			factor			0.30				
	0.60	0.60				10.90		0.95			10.40	
		10.90										
2	10.00	10.00	PVC ins 2.5mm² 2 core PVC									
	0.30	0.30	shthd cable 350/500V									
2	0.60	0.60	fixed to surface									
		10.90	supply	Y61.095/1	m	10.90	0.37			4.03		
			fix	Y61.095/3	m	10.90		3.18			34.66	
	2	2	Reconnection	Y61.1793	Ea	2		3.77			7.54	
										4.03	52.60	0.00
					Updating June 2009					0.62	9.94	0.00
							A	BDE	C			
							15.5%	18.9%	15.9%			
										4.65	62.54	0.00
					Contract % adjustment					-0.47	-6.25	0.00
							A	BDE	C			
							-10.0%	-10.0%	-10.0%			
										4.18	56.29	0.00
										Grand total		**60.47**

Measured Term Contract Order No:-	*EXAMPLE E6*
Location:-	
Work to be completed by:-	

Please execute the following:-

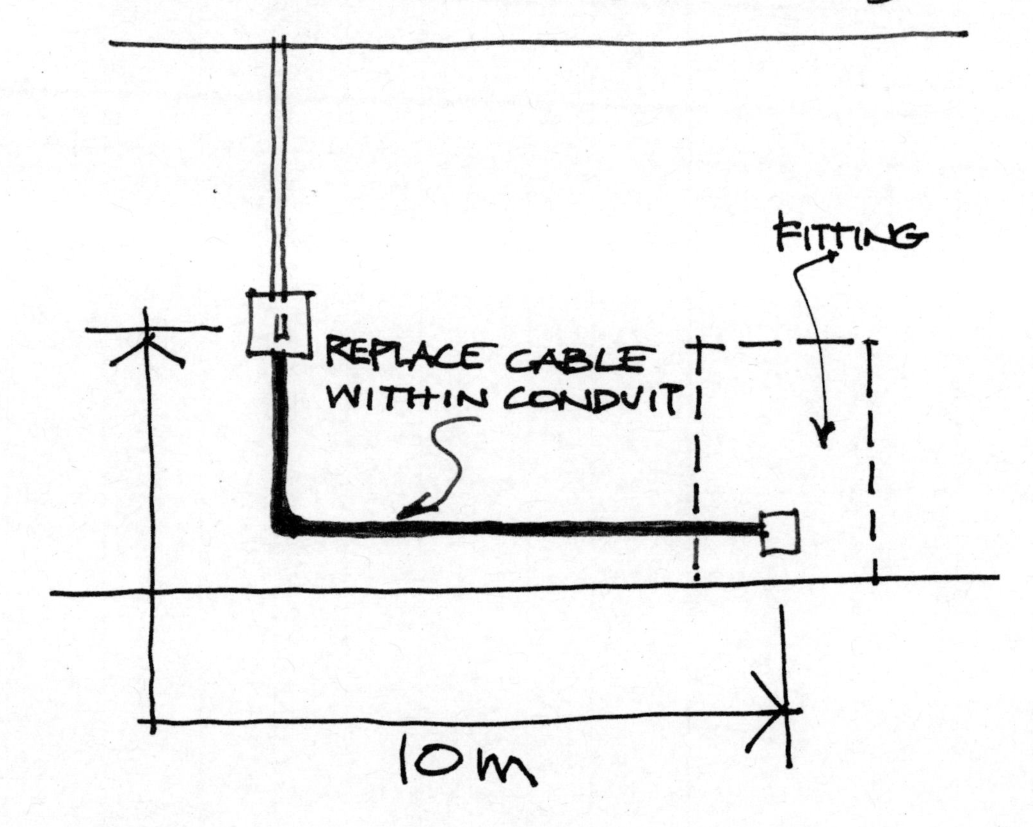

Signed for & on behalf of the PM

....

Date

....

Measured Term Contract Order No:-

ELECTRICAL 2006 - EXAMPLE 6

Measurement & Valuation

			Item ref	*Unit*	*Quantity*	*Rate* *A*	*BDE*	*C*	*Extension* *A*	*BDE*	*C*
10.00	10.00	Strip out extg cable	Y61.106/2	m			0.71				
0.30	0.30	from conduit		factor			0.30				
0.60	0.60				10.90		0.21			2.32	
	10.90										
10.00	10.00	PVC ins 2.5mm² non-shthd	Y61.106/1	m	10.90	0.17			1.85		
0.30	0.30	cable 450/750V drawn									
0.60	0.60										
	10.90										
		Ditto - fix	Y61.106/2	m	10.90		0.71			7.74	
2	2	Reconnection	Y61.1793	Ea	2		3.77			7.54	

	A	BDE	C
	1.85	17.60	0.00
Updating June 2009 (A 15.5%, BDE 18.9%, C 15.9%)	0.29	3.33	0.00
	2.14	20.93	0.00
Contract % adjustment (A -10.0%, BDE -10.0%, C -10.0%)	-0.21	-2.09	0.00
	1.93	18.84	0.00
Grand total			**20.77**

Measured Term Contract Order No:- *EXAMPLE E7*

Location:-

Work to be completed by:-

Please execute the following:-

Add new metalclad surface fixed switched socket outlet and 20 mm black enamel metal conduit to existing main (assume 2.00 m conduit)

Signed for & on behalf of the PM

....

Date

....

Measured Term Contract Order No:-

ELECTRICAL 2006 - EXAMPLE 7

Measurement & Valuation

				Item ref	*Unit*	*Quantity*	*Rate* A	*Rate* BDE	*Rate* C	*Extension* A	*Extension* BDE	*Extension* C
	2.00		Supply 20mm conduit	Y60.002/1	m	2.00	3.18			6.36		
		2.00	Fix only 20mm conduit	Y60.002/3	m	2.00		11.30			22.60	
	1		Conduit junc box - tee									
			supply	Y60.011/1	Ea	1	2.83			2.83		
			fix	Y60.011/5	Ea	1		3.53			3.53	
	1		Conduit junc box - lid									
			supply	Y60.063/1	Ea	1	0.38			0.38		
			fix	Y60.063/5	Ea	1		1.18			1.18	
	1		Switch socket outlet metalclad									
			supply	Y74.203/3	Ea	1	8.01			8.01		
			fix	Y74.203/5	Ea	1		10.36			10.36	
3	2.00	6.00	PVC ins 2.5mm² non-shthd									
3	0.30	0.90	cable 450/750V									
		6.90	supply	Y61.106/1	m	6.90	0.17			1.17		
			fix	Y61.106/2	m	6.90		0.71			4.90	
										18.75	42.57	0.00
					Updating June 2009					2.91	8.05	0.00
							A	BDE	C			
							15.5%	18.9%	15.9%			
										21.66	50.62	0.00
					Contract % adjustment					-2.17	-5.06	0.00
							A	BDE	C			
							-10.0%	-10.0%	-10.0%			
										19.49	45.56	0.00
										Grand total		**65.05**

Measured Term Contract Order No:- *EXAMPLE E8*

Location:-

Work to be completed by:-

Please execute the following:-

ADD 13A SOCKET OUTLET (1 GANG SP SW'D) TO EXISTING CIRCUIT ON OTHER SIDE OF WALL TO EXISTING RING. SURFACE MOUNTED INSTALLATION

ELEVATION

NEW 20mm CONDUIT & WIRING

EXTG INSTALLATION 20mm BLACK COND $2.5mm^2$ PVC INS CABLE NON-SHTHD, 450/750V.

NEW S/O SP SW'D METALCLAD SURFACE MOUNTED

NEW SMALL BOX MAL. IR WITH O'LAPPING LIGHT STEEL LID

1200

10 000

250

SECTION

EXTG RC WALL FAIR FACE BOTH SIDES

NEW S/O

ELBOW THRU WALL

NEW CONDUIT AND WIRE RISING FROM EXTG RING CIRCUIT AND TAKE THRU WALL

NB. ALL DIMS ARE IN mm.

Notes: Scrap extg wiring between two existing s/o's
Retain existing conduit

Signed for & on behalf of the PM

....

Date

....

Measured Term Contract Order No:-

ELECTRICAL 2006 - EXAMPLE 8

Measurement & Valuation

				Item ref	Unit	Quantity	Rate A	Rate BDE	Rate C	Extension A	Extension BDE	Extension C
3	10.00	30.00	Strip out extg cable	Y61.106/2	m			0.71				
6	1.20	7.20	from conduit		factor			0.30				
6	0.30	1.80				39.00		0.21			8.31	
		39.00										
	10.00		T/D extg 20ø conduit	Y60.002/3	m			11.30				
					factor			0.70				
						10.00		7.91			79.10	
	10.00		Fix only 20mm conduit	Y60.002/3	m	11.45		11.30			129.39	
	1.20											
	0.25											
		11.45										
	1.20		Supply 20mm conduit	Y60.002/1	m	1.45	3.18			4.61		
	0.25											
		1.45										
	1		Conduit junc box - tee supply	Y60.011/1	Ea	1	2.83			2.83		
			fix	Y60.011/5	Ea	1		3.53			3.53	
	1		Conduit junc box - lid supply	Y60.063/1	Ea	1	0.38			0.38		
			fix	Y60.063/5	Ea	1		1.18			1.18	
	1		Socket outlet - supply	Y74.203/3	Ea	1	8.01			8.01		
			fix	Y74.203/5	Ea	1		10.36			10.36	
3	10.00	30.00	PVC ins 2.5mm² non-shthd	Y61.106/1	m	49.50	0.17			8.42		
6	1.20	7.20	cable 450/750V - supply									
6	1.20	7.20										
6	0.25	1.50										
12	0.30	3.60	Ditto - fix	Y61.106/2	m	49.50		0.71			35.15	
		49.50										
	1		Cutting hole for conduit through 250mm RC wall	BW.125/1	Ea	1			49.90			49.90
	2		Make good fair face	BW.130/1	Ea	2			1.06			2.12
										24.25	267.02	52.02
				Updating June 2009			A 15.5%	BDE 18.9%	C 15.9%	3.76	50.47	8.27
										28.01	317.49	60.29
				Contract % adjustment			A -10.0%	BDE -10.0%	C -10.0%	-2.80	-31.75	-6.03
										25.21	285.74	54.26
										Grand total		**365.21**

Measured Term Contract Order No:-	*EXAMPLE M1*
Location:-	
Work to be completed by:-	

Please execute the following:-

EXTEND EXISTING 28mm COPPERTUBE 2m AND JOINT TO EXISTING

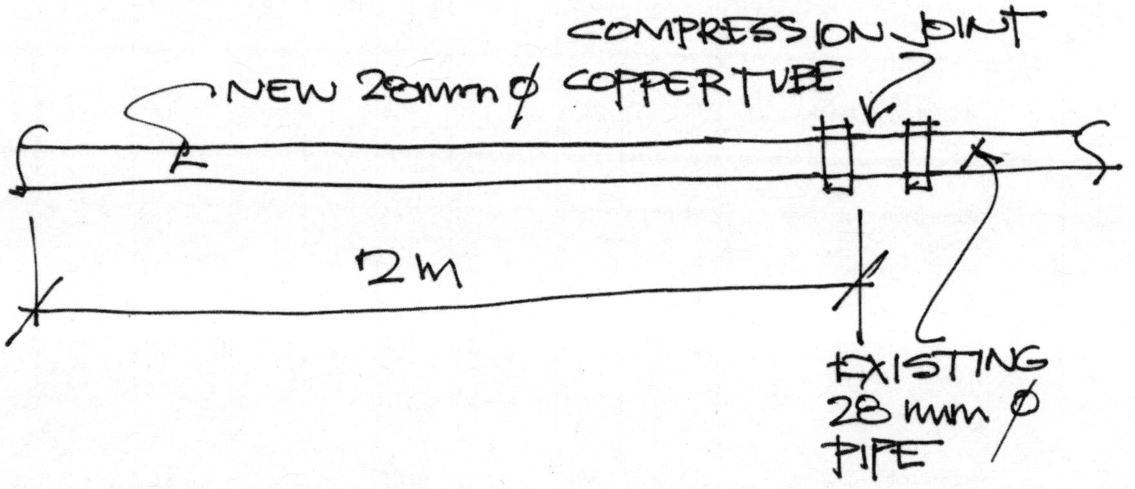

NEW 28mm Ø COPPERTUBE

COMPRESSION JOINT

2M

EXISTING 28 mm Ø PIPE

Signed for & on behalf of the PM

....

Date

....

Measured Term Contract Order No:-

MECHANICAL 2006 - EXAMPLE 1

Measurement & Valuation

				Item ref	*Unit*	*Quantity*	*Rate* A	*Rate* BDE	*Rate* C	*Extension* A	*Extension* BDE	*Extension* C
	2.00		28ø copper pipe / table X									
		2.00	Supply	Y10.235/1	m	2.00	3.59			7.18		
			Instal	Y10.235/4	m	2.00		4.20			8.40	
	1		Jointing pipe to existing									
		1	Instal	Y10.274/1	Ea	1		11.04			11.04	
										7.18	19.44	0.00
				Updating June 2009			A 19.7%	BDE 17.6%	C 15.9%	1.41	3.42	0.00
										8.59	22.86	0.00
				Contract % adjustment			A -10.0%	BDE -10.0%	C -10.0%	-0.86	-2.29	0.00
										7.73	20.57	0.00
										Grand total		**28.30**

Measured Term Contract Order No:- *EXAMPLE M2*

Location:-

Work to be completed by:-

Please execute the following:-

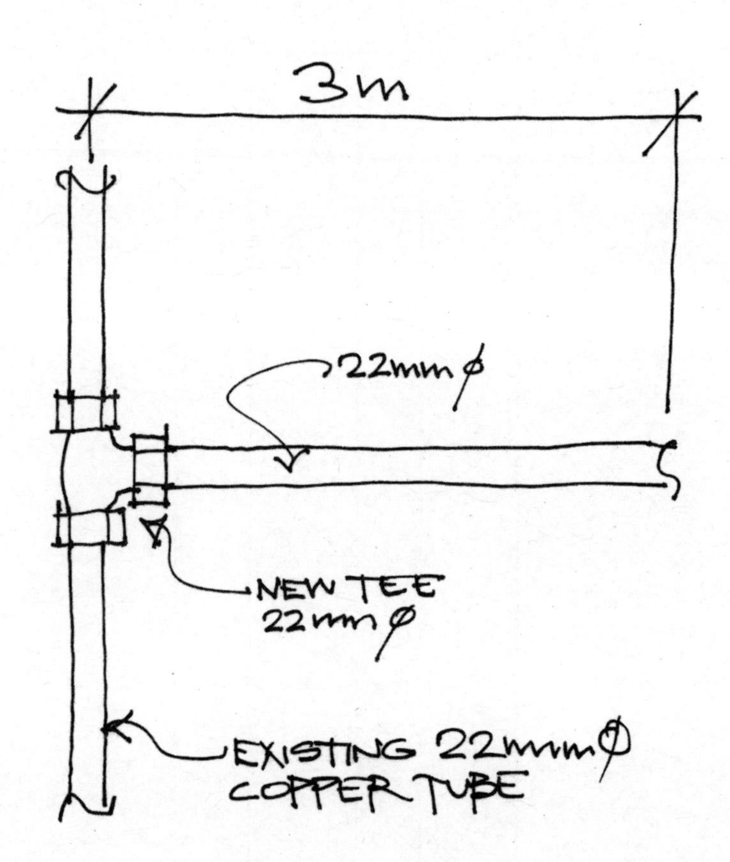

Signed for & on behalf of the PM

....

Date

....

Measured Term Contract Order No:-

MECHANICAL 2006 - EXAMPLE 2

Measurement & Valuation

				Item ref	Unit	Quantity	Rate A	Rate BDE	Rate C	Extension A	Extension BDE	Extension C
	3.00		22ø copper pipe / table X									
		3.00	Supply	Y10.234/1	m	3.00	2.74			8.22		
			Instal	Y10.234/4	m	3.00		3.99			11.97	
	1		22ø copper equal tee									
		1	Supply	Y10.260/2	nr	1	3.50			3.50		
			Instal	Y10.260/5	nr	1		14.29			14.29	
	1		Cutting into existing									
		1	Instal	Y10.281/1	Ea	1		18.08			18.08	
										11.72	44.34	0.00
				Updating June 2009						2.31	7.80	0.00
							A	BDE	C			
							19.7%	17.6%	15.9%			
										14.03	52.14	0.00
				Contract % adjustment						-1.40	-5.21	0.00
							A	BDE	C			
							-10.0%	-10.0%	-10.0%			
										12.63	46.93	0.00
										Grand total		**59.56**

Measured Term Contract Order No:- *EXAMPLE M3*

Location:-

Work to be completed by:-

Please execute the following:-

DISCONNECT EACH END AND TAKE DOWN 28mm COPPER TUBE AND STOP VALVE: SUPPLY AND FIX NEW COPPER TUBE AND COPPER DZR STOP VALVE WITH COMPRESSION JOINTS (ASSUME 5 METRES TOTAL LENGTH)

28mmØ COPPER TUBE

COMPRESSION JOINT

STOP VALVE

5000

Signed for & on behalf of the PM

....

Date

....

Measured Term Contract Order No:-

MECHANICAL 2006 - EXAMPLE 3

Measurement & Valuation

			Item ref	*Unit*	*Quantity*	*Rate* *A*	*Rate* *BDE*	*Rate* *C*	*Extension* *A*	*Extension* *BDE*	*Extension* *C*
5.00		Take down extg 28ø cu pipe	Y10.235/4	m			4.20				
	5.00			factor			0.20				
					5.00		0.84			4.20	
5.00		Supply									
	5.00	28ø copper pipe / table X	Y10.235/1	m	5.00	3.59			17.95		
5.00		Instal									
	5.00	28ø copper pipe / table X	Y10.235/4	m	5.00		4.20			21.00	
2		Instal									
	2	Jointing pipe to existing	Y10.274/1	Ea	2		11.04			22.08	
1		Supply									
	1	Cu alloy stop vlv / 25ø / compression, DZR	Y11.003/4	Ea	1	24.53			24.53		
1		Instal									
	1	Cu alloy stop vlv / 25ø	Y11.003/5	Ea	1		11.35			11.35	
									42.48	58.63	0.00
			Updating June 2009						8.37	10.32	0.00
						A	BDE	C			
						19.7%	17.6%	15.9%			
									50.85	68.95	0.00
			Contract % adjustment						-5.09	-6.90	0.00
						A	BDE	C			
						-10.0%	-10.0%	-10.0%			
									45.76	62.05	0.00
									Grand total		**107.81**

Measured Term Contract Order No:- *EXAMPLE M4*

Location:-

Work to be completed by:-

Please execute the following:-

TAKE DOWN 28mm COPPER TUBE AND STOP VALVE: SUPPLY AND FIX NEW COPPER TUBE AND REUSE EXISTING VALVE (ASSUME 5m TOTAL LENGTH)

CEILING

300

28mm ø COPPER TUBE

5m

REUSED STOP VALVE

Signed for & on behalf of the PM

....

Date

....

Measured Term Contract Order No:-

MECHANICAL 2006 - EXAMPLE 4

Measurement & Valuation

				Item ref	*Unit*	*Quantity*	*Rate A*	*Rate BDE*	*Rate C*	*Extension A*	*Extension BDE*	*Extension C*
	5.00		Take down extg 28ø cu pipe	Y10.235/4	m			4.20				
		5.00			factor			0.20				
						5.00		0.84			4.20	
	5.00		Supply									
		5.00	28ø copper pipe / table X	Y10.235/1	m	5.00	3.59			17.95		
	5.00		Instal									
		5.00	28ø copper pipe / table X	Y10.235/4	m	5.00		4.20			21.00	
	2		Instal									
		2	Jointing pipe to existing	Y10.274/1	Ea	2		11.04			22.08	
	1		Take down extg									
		1	Cu alloy stop vlv / 25ø /	Y11.003/5	Ea			11.35				
					factor			0.75				
						1		8.51			8.51	
	1		Instal									
		1	Cu alloy stop vlv / 25ø	Y11.003/5	Ea	1		11.35			11.35	

	A	BDE	C
	17.95	67.14	0.00
Updating June 2009 (A 19.7%, BDE 17.6%, C 15.9%)	3.54	11.82	0.00
	21.49	78.96	0.00
Contract % adjustment (A -10.0%, BDE -10.0%, C -10.0%)	-2.15	-7.90	0.00
	19.34	71.06	0.00
Grand total			**90.40**

Measured Term Contract Order No:- *EXAMPLE M5*

Location:-

Work to be completed by:-

Please execute the following:-

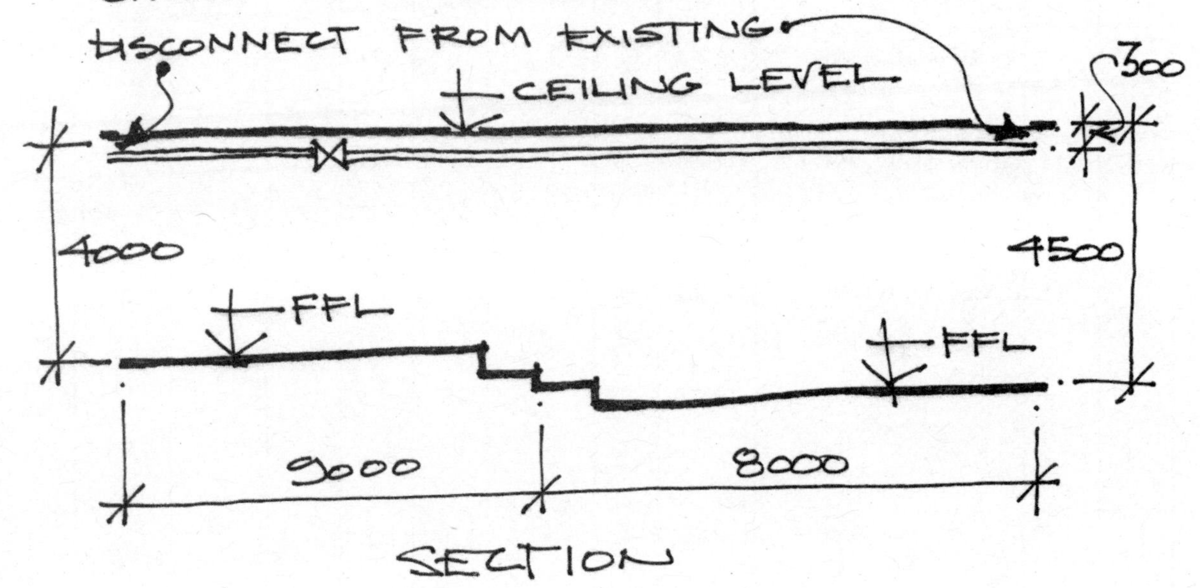

Signed for & on behalf of the PM

....

Date

....

Measured Term Contract Order No:-

MECHANICAL 2006 - EXAMPLE 5

Measurement & Valuation

			Item ref	*Unit*	*Quantity*	*Rate A*	*Rate BDE*	*Rate C*	*Extension A*	*Extension BDE*	*Extension C*
9.00		T/D extg 28ø cu pipe	Y10.235/4	m			4.20				
	9.00			factor			0.20				
					9.00		0.84			7.56	
8.00		T/D extg 28ø cu pipe/	Y10.235/4	m			4.20				
	8.00	4-7m above firm base		factor			0.20				
				factor			1.25				
					8.00		1.05			8.40	
9.00		Supply									
8.00		28ø copper pipe / table X	Y10.235/1	m	17.00	3.59			61.03		
	17.00										
		Instal									
9.00		28ø copper pipe / table X	Y10.235/4	m	9.00		4.20			37.80	
	9.00										
		Instal									
8.00		28ø copper pipe / table X/	Y10.235/4	m			4.20				
	8.00	4-7m above firm base		factor			1.25				
					8.00		5.25			42.00	
		Instal									
1		Jointing pipe to existing	Y10.274/1	Ea	1		11.04			11.04	
	1										
		Instal									
1		Jointing pipe to existing/	Y10.274/1	Ea			11.04				
	1	4-7m above firm base		factor			1.25				
					1		13.80			13.80	
1		Supply									
	1	Cu alloy check vlv / 25ø /	Y11.258/1	Ea	1	51.31			51.31		
		compression, double									
		Instal									
1		Cu alloy check vlv / 25ø	Y11.258/3	Ea	1		11.35			11.35	
	1										
									112.34	131.95	0.00
				Updating June 2009					22.13	23.22	0.00
						A	BDE	C			
						19.7%	17.6%	15.9%			
									134.47	155.17	0.00
				Contract % adjustment					-13.45	-15.52	0.00
						A	BDE	C			
						-10.0%	-10.0%	-10.0%			
									121.02	139.65	0.00
									Grand total		**260.67**

Measured Term Contract Order No:- *EXAMPLE M6*

Location:-

Work to be completed by:-

Please execute the following:-

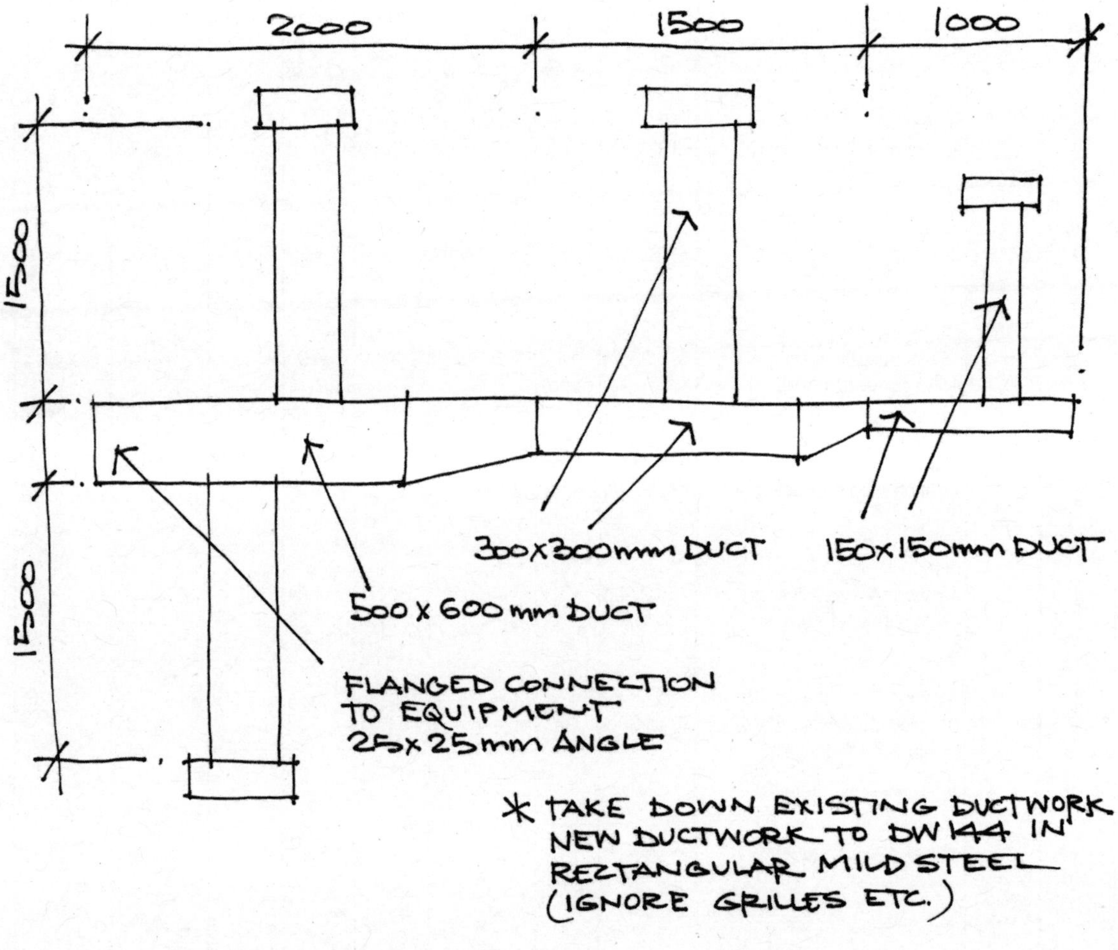

2000 1500 1000

1500

1500

300 x 300 mm DUCT

150 x 150 mm DUCT

500 x 600 mm DUCT

FLANGED CONNECTION
TO EQUIPMENT
25 x 25 mm ANGLE

* TAKE DOWN EXISTING DUCTWORK
NEW DUCTWORK TO DW 144 IN
RECTANGULAR MILD STEEL
(IGNORE GRILLES ETC.)

Signed for & on behalf of the PM

....

Date

....

Measured Term Contract Order No:-

MECHANICAL 2006 - EXAMPLE 6

Measurement & Valuation

				Item ref	*Unit*	*Quantity*	*Rate* A	*Rate* BDE	*Rate* C	*Extension* A	*Extension* BDE	*Extension* C
	2.00		500 x 500 galv. m.s.	Y30.015/2	m			26.93				
		2.00	std gauge duct		factor			0.75				
			Take down extg			2.00		20.20			40.40	
			Supply	Y30.015/1	m	2.00	34.73			69.46		
			Instal	Y30.015/2	m	2.00		26.93			53.86	
	1		E.O. for reducing piece									
		1	Supply	Y30.073/3	Ea	1	43.06			43.06		
			Instal	Y30.073/5	Ea	1		26.70			26.70	
	1.50	1.50	300 x 300 galv. m.s.	Y30.009/2	m			21.44				
3	1.50	4.50	std gauge duct		factor			0.75				
		6.00	Take down extg			6.00		16.08			96.48	
			Supply	Y30.009/1	m	6.00	20.29			121.74		
			Instal	Y30.009/2	m	6.00		21.44			128.64	
	1		E.O. for reducing piece									
		1	Supply	Y30.067/3	Ea	1	26.67			26.67		
			Instal	Y30.067/5	Ea	1		21.44			21.44	
	1.00		150 x 150 galv. m.s.	Y30.002/2	m			17.64				
	1.00	2.00	std gauge duct		factor			0.75				
			Take down extg			2.00		13.23			26.46	
			Supply	Y30.002/1	m	2.00	15.55			31.10		
			Instal	Y30.002/2	m	2.00		17.64			35.28	
	1		E.O. for blanking off plate									
		1	Supply	Y30.032/1	Ea	1	10.99			10.99		
			Instal	Y30.032/5	Ea	1		8.83			8.83	
4	0.50	2.00	Flanged & bolted (to equipment									
12	0.30	3.60	joint (300mm joint									
4	0.15	0.60	Supply (150mm joint	Y30.029/1	m	6.80	17.16			116.69		
4	0.15	0.60	Instal (150mm blank end	Y30.029/3	m	6.80		9.46			64.33	
		6.80										
										419.71	502.42	0.00
				Updating June 2009						82.68	88.43	0.00
							A 19.7%	BDE 17.6%	C 15.9%			
										502.39	590.85	0.00
				Contract % adjustment						-50.24	-59.09	0.00
							A -10.0%	BDE -10.0%	C -10.0%			
										452.15	531.76	0.00
										Grand total		**983.91**

10 Worked Examples – notes on solutions

Example BW1: Cutting opening for door in one brick wall

"6mm mild steel bar reinforcement"

Note that reinforcement is per kg.

Example BW2: Brick up opening in existing wall

"Bonding end one brick thick wall"

The new brickwork must be bonded to the existing brickwork.

"Extra over for facings"

Brickwork would be "composite work" not facing bricks throughout.

Example BW3: Partition of 100 mm thick blockwork

"19 x 75mm (finished size) softwood skirting"

The measurement rules on Page 403 state "Measure the finished size" and "Multiply the width by the depth without deduction". As the size is shown as "finished", no allowance for planing (usually about 3mm per wrought face) is made.

Finished size 19 x 75mm = 1425 mm²
P20.014/1 is the rate for the first 500 mm² = £2.97
P20.014/2 = (1425 - 500)/100 = 9.25 x £0.33 = £3.05

Example BW4: Repairing defective softwood flooring

"Floor member"

The rates for taking down the existing timber, new floor members and timber preservation are all stated as "per 10 000 mm² of sectional area". Floor members in the example are 50 x 100 mm = 5000 mm².

Example BW5: Timber stud partition with plasterboard

“Partition member”

As “Floor member” in Example BW04.

“19 x 75mm (finished size) softwood skirting”

As “Skirting” in Example BW03.

Example BW6: Door frame

“Door frame hardwood 100 x 75mm (nominal) teak”

The measurement rules on Page 254 state that the preambles to Section L10 apply equally to Section L20 and on Page 237 state “Measure the finished size” and “Multiply the width by the depth without deduction”. As the size is shown as nominal, an allowance for planing (usually about 3mm per wrought face) is made.

Finished size 94 x 69mm = 6486mm^2
L20.103/3 is the rate for the first 500mm^2 = £2.09
L20.103/4 = (6486 - 500)/100 = 60 x £0.27 = £16.20
L10.001 is the multiplying factor for timber ordered to be selected and kept clean for clear finishing.
K20.007 is the multiplying factor for teak.

“Rebate”

L20.110/2 is the rate for a rebate in hardwood.
K20.007 is the multiplying factor for teak.

“Plugging”

P20.119 is the rate for a single row of plugging to masonry.

“Screwing and pelleting”

P20.116/2 is the rate for fixing with brass screws to hardwood.
P20.118/2 is the rate for pelleting.
K20.007 is the multiplying factor for teak.

Example BW7: Repairing plaster and lath ceiling

"In repairs 1 to 5 m²"

"In repairs" is deemed to include removing existing work, new work to match existing, preparatory work and making good, jointing new to existing and work of any width.

For areas exceeding the size ranges stated in the "ADD where in repairs" descriptions, each item of replacement work is measured separately.

Example BW8: Repairing plaster to concrete ceiling

"Hack off plaster from ceiling" and "Jointing to existing"

The area exceeds the size ranges stated in M20.008, therefore each item of replacement work is measured separately.

Example BW9: Patch repair plaster to brick wall

"In repairs 1 to 5 m²"

As Example BW07.

Example BW10: Repairing plaster to brick wall

"Hack off plaster from wall", "Form key" and "Jointing to existing"

As Example BW08.

Example BW11: Repairing wall tiling

"Render internal wall exceeding 300mm wide, 2 coats on brickwork" and "152 x 152 x 5.5mm white glazed wall tiles"

Both descriptions are within the size range of "In repairs" and are deemed to include removing existing work, new work to match existing, preparatory work and making good, jointing new to existing and work of any width.

Example BW12: Repairing bitumen macadam

Examples showing the difference in cost between the laying of an area 10 to 25m², an area 10m x 5m, and 250 patches of 2m² and the effect of the multiplier factor Q22.002 and the amount of jointing to existing. (see also "Frequently asked questions").

Example BW13: Repairing clay roof tiles, battens and felt

"Fixing only plain tiles"

The cost of tiles can vary considerably. Therefore, tile rates are for "fixing only" and do not include for the cost of tiles (but do include fixing materials).

The number of tiles per m² will vary according to the size, gauge and lap of the tile.

Example BW14: Patch repair three layer roofing felt

"In repairs 5 to 10 m²"

"In repairs" is deemed to include removing existing work, new work to match existing, preparatory work and making good, jointing new to existing and work of any width.

Note that it is applied to each layer.

Example BW15: Re-roofing with three layer roofing felt

"Skirting", "kerb" and "turn down at abutment"

These are all measured separately in metres.

Example BW16: Replace sink

"Take out sink"

Note that, in the case of sanitary appliances, the "Take out" rates are in Section N13 not C90.

"42 mm waste"

All pipework and associated fittings is measured and valued in accordance with Section R11: Foul drainage above ground.

Example BW17: Replace basin

"15 mm copper supply"

Waste pipework and associated fittings is measured and valued in accordance with Section R11: Foul drainage above ground.

Supply pipework and associated fittings is measured and valued in accordance with Section S10: Cold water.

Example BW18: Replace WC

As Example BW17.

Example BW19: Cold water storage tank

"Plastics cistern 227 litre, rectangular"

Note that cover, ball valve and perforations for pipes and all measured separately.

Example BW20: Replace PVC gutter and downpipe

"PVC gutter and downpipe"

Measure the length of gutters and downpipes over all fittings. Fittings are measured as "Extra over".

Example BW21: Replace length of defective drain pipework

"Connect new pipe to 150 mm existing"

For new work, joints in the running length including cutting and jointing to fittings is deemed to be included. Where connecting new pipe to existing pipe end, however, R12.118 or 119 are required.

Example BW22: Manhole

"Excavating pit starting at ground level maximum depth not exceeding 1.00m"

R12.250/1 is the rate for excavating a pit for the construction of a manhole. Note that backfilling, compacting and disposal of surplus excavated materials off site are deemed to be included.

"Disposal excavated material from site"

Disposal of surplus excavated materials off site is deemed to be included in R12.250/1.

"Earthwork support, distance between opposing faces not exceeding 2.00m maximum depth not exceeding 1.00m"

R12.252/1 is the rate for earthwork support. The maximum depth of excavation does not exceed 1.00m.

"1B wall in Engineering bricks in manhole"

R12.301/2 is the rate for wall one brick thick in Engineering bricks in manhole. The length is the "mean girth".

The cost of bricks can vary considerably. Therefore, brickwork rates are for "laying only" and do not include for the cost of bricks (but do include the cost of mortar).

The number of bricks can be calculated from the Tables in Section F10. In this case, for a one brick wall it should be 128 bricks per m². Note that the number of bricks includes an allowance for waste.

"150 mm half round channel 0.90 m girth"

R12.315/2 is the first 600 mm and R12.316/2 is for each 150 mm of additional length (0.90 - 0.60 = 0.30 / 0.15 = 2)

Example BW23: Strip foundation (common errors)

"Excavating trench starting at ground level not exceeding 2.00m deep over 0.30m wide"

The correct way of arriving at the length of the trench is by taking the "mean girth". This is calculated by taking half of the width of the trench from each end of all four sides (4 x 2 x 0.5 x 0.825m) and subtracting the result from the outer perimeter (20.000m).

The trench is not exceeding 1.00m deep (Item D20.0**/2 not D20.0**/3).

Trench "over 0.30m wide" is Item D20.014 not D20.013.

"Disposal of excavated material on site 100m and depositing"

D20.043 includes the first 25m distance.

The quantity of excavated material disposal needs to be reduced by the amount backfilled.

"Extra over for excavating below ground water level"

Only 0.40m of the excavation is below ground water level.

Pumping is not allowable. Excavating below ground water level includes for keeping excavations free of ground water (see "Rates for the following include" under EXCAVATING).

"Backfilling"

The correct way of arriving at the length of the trench is by taking the "mean girth" of the trench outside the brickwork in foundations for filling with earth from excavations and inside the brickwork in foundations for filling with imported hardcore.

"Earthwork support"

The support has a maximum depth not exceeding 1.00m (Item D20.032/1 not D20.032/2).

"Concrete in foundations"

The concrete mix on the Order is 15N/mm². The Rate for plain concrete in foundations Item E10.005/1 is for 20N/mm² and should be reduced by Item E10.029/1.

"Brickwork"

The bricks shown on the Order are common bricks not engineering bricks (Item F10.001/2 not F10.063/2). Item F10.001 is for common brickwork in cement lime mortar, therefore F10.015 cement mortar should be added.

The cost of bricks can vary considerably. Therefore, common brickwork rates are for "laying only" and do not include for the cost of bricks (but do include the cost of mortar).

The number of bricks can be calculated from the Table at the beginning of the common brickwork section of the Schedule. There are similar Tables at the beginning of the sections for facing bricks and engineering bricks. In this case, for a one brick wall it should be 128 bricks per m^2. Note that the number of bricks includes an allowance for waste.

In order to obtain competition, tenderers submit a contract percentage on or off the Schedule of Rates to reflect their assessment of all additional factors (such as overheads, market conditions, etc). The Contract percentage is subsequently applied to all work priced against the Schedule of Rates. Depending on the contract conditions, a second adjustment is sometimes made - to update the pricing level of the Schedule of Rates to the current month, using published percentages. It is essential that this percentage is applied at the appropriate place in the calculation. Invoices (for the supply of bricks, for example) will be at the base date of the work carried out and, therefore, should be added after the application of the contract and updating percentages.

Example BW24: Excavation for pit

"Excavating trench starting at ground level not exceeding 2.00m deep over 0.30m wide"

The first part of the excavation can be taken as a trench 5.00m long, 1.00m wide and 1.5m deep.

"Excavating pit starting 1.50m below ground level not exceeding 2.00m deep"

The second part of the excavation can be taken as a pit 2.00m long, 1.00m wide and 1.5m deep. This second excavation starts "over 0.25m below ground level" and will require the addition of item D20.017 x 1.50 = £5.13.

"Extra over for excavating below ground water level"

The excavation of the pit is also "below ground water level" and will require the addition of item D20.018.

"Disposal of excavated material on site 350m distance"

Excavation will require the disposal of excavated material either on or off site. The Order states "Deposit excavated material on site - 350m from site of excavation". Item D20.043 includes the first 25m of removal distance and therefore the rate for the remaining distance is calculated by D20.044 x 325/50 = £3.32.

"Levelling bottom of excavation"

It is usual to treat the surface of excavations, particularly where concrete will be poured on to it.

"Extra over for excavating next to service"
"Extra over for excavating around service crossing excavation"

Where an existing service (pipe, cable, etc.) is to be retained, additional precautions are required.

"Earthwork support distance between opposing faces not exceeding 2.00m, maximum depth not exceeding 2.00m"

Earthwork support includes the use of timber planking and strutting, plywood trench sheeting and light steel trench sheeting.

Earthwork support maximum depth not exceeding 2.00m is the shallow sides and shallow end.

"Earthwork support distance between opposing faces not exceeding 2.00m, maximum depth not exceeding 4.00m below ground water level"

Earthwork support maximum depth not exceeding 4.00m is the deep sides, deep end and other side of pit.

"Add for support next roadway maximum depth not exceeding 2.00m"
"Add for support next roadway maximum depth not exceeding 4.00m"

Where earthwork support is required to protect any adjacent roadway, D20.038 is measured. A guide is where the horizontal distance from the face supported to the edge of a roadway or footpath is less than the depth of the excavated face below the roadway or footpath (SMM7 D20 D6).

Example BW25: Suspended reinforced concrete slab

"Slab reinforced 150 to 450mm thick 30N/mm² 10mm aggregate ordinary cement"

The dimensions for the slab are the overall length x overall width x thickness plus the length x thickness x depth of the thicker of the two attached beams (see below for other beam).

"Attached deep beam reinforced 150 to 450mm thick 30N/mm² 10mm aggregate ordinary cement"

Deep beams are those whose depth (measured below the slab where attached) is greater than three times their width (See E10 Definition of terms). The thinner of the two attached beams is 200mm wide and 700mm deep and is therefore a deep beam. The other is 300mm wide and 700mm deep and is therefore not a deep beam and should be included in the measurement of the slab.

"Formwork to soffit of slab 300mm thick 3.00 to 4.50m high with plain finish"

E20.008 is the rate for soffit of slab not exceeding 200mm thick with basic finish.
E20.010 is the rate for the additional 100mm of thickness.
E20.045 is the multiplying factor for plain finish
E20.041 is the rate for the height adjustment for soffits

"Formwork to edge of suspended slab 250-500mm high with plain finish"

E20.003/3 is the rate for formwork to edge of suspended slab 250-500mm high with basic finish. The dimensions are for the two sides without attached beams.
E20.045 is the multiplying factor for plain finish

"Formwork to attached beam plain finish"

Formwork to edges of suspended slabs associated with attached beams at slab perimeters is included with the measurement of the formwork to such beams. E20.016 is the rate for attached beam with basic finish. The dimensions are the length of the beams x the height of the beams (for both the inside and outside surface of the beams) plus the soffits of the two beams. E20.045 is the multiplying factor for plain finish.

"Fix Only Reinforcement"

Rates for Fixing Only reinforcement are shown in column 3 and 4 of items E30.001 to 009, and column 2 of items E30.010 to 019.

"Float finish"

E41.002 is the rate for floating worked finish.

Example E1: Replace fluorescent lamps

"Taking down diffuser"

Y73.010/7 is the Rate B for installation, 0.70 is the multiplying factor for the Rate E dismantling for re-use.

"Taking down existing lamps"

Y73.967/1 is the Rate B for installation, 0.40 is the multiplying factor for the Rate D dismantling for scrap.

Example E2: Renew batten luminaire

"Dismantle and remove existing luminaire "

Y73.006/9 is the Rate B for installation, 0.40 is the multiplying factor for the Rate D dismantling for scrap.

Example E3: Renew modular fluorescent luminaire

"Re-fix one existing and supply and fix one new"

Y73.030/5 is the Rate B for installation, 0.40 is the multiplying factor for the Rate D dismantling for scrap (in preparation for fixing new), 0.70 is the multiplying factor for the Rate E dismantling for re-use (in preparation for re-fixing existing).

Example E4: Replace bathroom fan

"Dismantle and remove domestic bathroom fan"

Y41.001/4 is the Rate B for installation, 0.75 is the multiplying factor for the Rate D dismantling for scrap.

Note that this is different to the Examples above. Check the first page of each Section for Rate D and E.

Example E5: Replace cable fixed to surface

"Strip out existing cable between fitting and equipment"

Y61.095/3 is the rate for installing the cable fixed to any background (Rate B). Dismantling and removing for scrap including disposal (Rate D) is the Rate B multiplied by 0.30.

The length of stripped out cable is calculated by taking the distance between the fitting and equipment without any allowance for sag. To this should be added 0.30m per cable entering the fitting and 0.60m per cable entering the equipment.

"Reconnection"

It is the items to which cable ends and tails are fixed that include for connection in their Rate B, not the cable itself. If it is the cable itself that is being removed and the fitting, accessory, etc. that remains, then items Y61.1793 to 1800 (connections to existing equipment), would be used.

Example E6: Replace cable in conduit

"Strip out existing cable between fitting and equipment"

Y61.106/2 is the rate for installing the cable drawn through conduit (Rate B). Dismantling and removing for scrap including disposal (Rate D) is the Rate B multiplied by 0.30.

Example E7: Socket outlet on existing main

"Conduit junction box"

Measure the length of conduits over all conduit boxes, conduit fittings, etc. Fittings are measured as "Extra over".

Example E8: Electrical installation

"Strip out existing cable"

Y61.106/2 is the rate for installing the PVC insulated 2.5mm² non-sheathed 450/750V cable (Rate B). Dismantling and removing for scrap including disposal (Rate D) is the Rate B multiplied by 0.30.

The length of stripped out cable is calculated by taking the distance between the two existing socket outlets x 3 (three cables) and adding the height to the two socket outlets x 2 per socket (to the socket and return) x 3 (three cables). To this should be added 0.30m per cable entering the socket.

"Take down existing 20mm diameter conduit"

Y60.002/3 is the rate for installing the 20mm conduit (Rate B). Dismantling and removing for scrap including disposal (Rate D) is the Rate B multiplied by 0.25.

"PVC insulated 2.5mm^2 non-sheathed 450/750V cable"

Y61.106/1 is the rate for supplying the PVC insulated 2.5mm^2 non-sheathed 450/750V cable (Rate A).

The length of new cable is calculated by taking the distance between the two existing socket outlets x 3 (three cables) and adding the height to the two existing socket outlets x 1 per socket and the new socket outlet (to the socket and return) x 3 (three cables). To this should be added 0.30m per cable entering the socket.

"Cutting hole for conduit through 250mm reinforced concrete wall"

BW.125/1 is the rate for cutting hole through reinforced concrete per 25mm of thickness.

Example M1: Extending existing copper tubing

"Jointing pipe to existing"

For new work, joints in the running length of copper tubing are deemed to be included. Where jointing new pipe to existing pipe end, however, Y10.268 to 280 are required.

Example M2: Inserting fitting into existing copper tubing

"Cutting into existing"

Where cutting into existing copper tubing for insertion of a new fitting, Y10.281 to 283 are required.

Example M3: Replacing copper tubing and stop valve

"Take down existing 28 mm diameter copper pipe"

Dismantling and removing for scrap (Rate D) is measured over all fittings, but not the fittings. It is assumed that, in scrapping, the stop valve would not be dismantled.

Example M4: Replacing copper tubing and re-using existing stop valve

"Take down existing 28 mm diameter copper pipe"

Dismantling and removing for re-use (Rate E) is measured over all fittings. If the fittings, in this case the stop valve, are to be re-used, then dismantling and removing for re-use (Rate E) is measured.

Example M5: Pipework renewal

"Take down existing 28mm diameter copper pipe"

Y10.235/4 is the rate for installing the copper pipe (Rate B). Dismantling and removing for scrap including disposal (Rate D) is the Rate B multiplied by 0.20 (see Page 133).

When using Rate D, dismantling and removal for scrap including disposal, it is assumed that pipes will be cut into convenient lengths without removing every fitting (tee, bend, etc.). Rate D would be applied to the length of the pipe measured over all fittings, but not to each actual fitting.

When using Rate E, dismantling and removal for re-use, then each fitting to be re-used would be measured.

"Take down existing 28mm diameter copper pipe 4-7m above firm base"

The multiplying factors for carrying out work exceeding 4m from a firm base are shown on page 4 of the Mechanical Services Schedule.

Example M6: Ductwork renewal

"Galvanised mild steel ductwork"

Installing ductwork (Rate B) is measured along the centre line over all fittings and branches. Installing fittings (Rate B) is measured as "Extra over".

"Flanged and bolted joint"

Flanged and bolted joints are measured per metre of duct girth to flanges on equipment, to receive blanking plates and to cross joints.

11 Frequently asked questions

Q. Do the rates include for labour and materials?

A. All rates include for labour, materials and wastage, unless otherwise stated in the headings. Rates under "Fixing only doorsets" on page 262 of the Building Works Schedule, for example, do not include for the doorsets. "Fixing only" or "laying only" rates only include for labour and fixing or jointing materials (and wastage on those fixing or jointing materials).

However, in full, the General Directions, page 4, state that rates throughout include:

"Fixing only or laying only:

For fixing only or laying only new items or items previously set aside for re-use.

For taking delivery, storing and sending back returnable packings.

For obtaining from stack or other place of storage on site.

For handling, loading, unloading, protecting, transporting to the site of the work, hoisting, lowering, assembling and fixing complete.

For all fixing or jointing materials required."

Q. Do the rates include for disposal and removal from site?

A. Rates include for disposal and removal from site. The General Directions, page 4, state that rates throughout include:

"For getting out from the interior to the exterior of the building and lowering or hoisting to ground level as necessary by means of barrows and wheeling gangways, chutes, bagging, bucketing or hoists.

For removing off the site to an approved place of disposal including payment of landfill tax and any other charges in connection therewith."

Q. What if the particular circumstances of the carrying out of some work results in the cost to the Contractor exceeding the value as measured and valued?

A. The General Directions, page 3, state that rates throughout include:

"For carrying out work in any circumstances, unless otherwise stated. It is assumed that work carried out in disadvantageous circumstances (e.g. work in occupied buildings, alterations, repairs or extensions) will be offset by work carried out in advantageous circumstances."

The price paid is a classic case of 'swings and roundabouts'. If a Contractor has to replace all the rainwater goods on a housing estate the Schedule of Rates will reimburse him handsomely - on the other hand if he is required to travel 50 miles to replace a pane of glass he will not be so fortunate. It is important therefore, to order all work covered by the Schedule of Rates from the Contractor - not just the low value and awkward work. The Contractor will have tendered on the basis of all relevant work with a spread of value.

Q. Do the rates include for plant? Do the rates for disposal of waste materials include for skips and the like?

A. Rates are deemed to include for all plant including skips, (but see below for separate rules for scaffolding). The General Directions, page 1, state:

"The Rates in this Schedule are all inclusive and, as such, include for Preliminaries."

"When assessing any adjustment to be tendered to the level of the Rates, consideration must be given to any unique preliminary items . . . . . . Plant, tools, vehicles and transport."

Q. Do the rates include for scaffolding?

A. Rates are deemed to include for all scaffolding up to 4.50m. The General Directions, page 1, state that rates throughout include:

"For providing scaffolding or working platforms not exceeding 4.50 m above the base of any scaffolding."

A44: Temporary Works, page 9, states:

"Scaffolding or working platforms not exceeding 4.50m high: Are included in the Rates throughout the Schedule."

Measure the height of scaffolding to be valued, using the scaffolding rates on page 10, from 4.50 m above the base of the scaffold to the highest working platform. See method of measurement on page 9 for further details.

Q. If a Sub-Contractor or Supplier invoice shows a unit price and a discount of say 10%, is the Contractor obliged to declare

the discount in its entirety or can he withhold 2.5% and subsequently give back the remaining 7.5%? In other words should the discount be 10% or 7.5%?

A. The General Directions, page 2, state that:

"In the case of work ordered to be placed with a Sub-Contractor or Supplier, the Contractor, unless the Conditions of Contract state otherwise, will be reimbursed the net agreed amount of their account (after the deduction of all discounts obtainable for cash, insofar as they exceed 2½ per cent, and of all trade discounts, rebates and allowances) with the addition of 5 per cent to cover profit and all other liabilities. Such reimbursement shall not be subject to further adjustment."

The answer depends on the terms of the invoice and the nature of the discount. Invoices usually include an allowance for discount for cash of 2½% for payment within 30 days. If this is the case, then the contractor deducts 2½% from the final invoiced amount (the amount after all other discounts) when he pays. The phrase "insofar as they exceed 2½%" prevents the Contractor from losing his discount for cash of 2½%.

In this example, if the Contractor is able to obtain a discount for cash of 2½% from the sum remaining after the deduction of 10% discount for cash, then, the full 10% should be deducted. If only 7½% is deducted he will be getting the 2½% twice. However, if the invoice does not allow him to take a discount for cash of 2½%, then 10% discount for cash would be deducted but 1/39th added (the amount equivalent to 2½% if deducted).

The only way of knowing which method to use is to find out what the Contractor is actually paying. A simple solution would be to deduct the full 10% cash discount until the Contractor provides evidence that he is not receiving any further cash discounts and, when the evidence is provided, add 1/39th.

It should be noted that the above applies only to cash discounts. All trade discounts, rebates and allowances are deducted. Therefore, if the 10% is a trade discount not a cash discount and the invoice does not allow him to take a discount for cash of 2½%, then 1/39th is not added. He cannot retain what was not there in the first place.

Q. Section A: General Directions, Rates throughout include, refers to "For square cutting". Would this include saw cutting bitumen macadam, etc?

A. A clear distinction should be made between measurement rules and method of working. All "square cutting" is deemed to be included and this is a measurement rule. The method of achieving the square cutting is irrelevant.

Q. Do the rates include for waste? Do "Fixing only" rates include for waste?

A. All items in the Schedule that contain materials include an element of waste. The amount will vary according to the item and the material. "Fixing only" rates only include for labour and fixing or jointing materials (and wastage on those fixing or jointing materials) but not the material itself and, therefore, cannot include for wastage on that material.

Q. Some rates are described as "ADD where in repairs". What does this mean?

A. "In repairs" is deemed to include removing existing work and jointing new to existing.

For areas exceeding the size ranges stated in the "ADD where in repairs" descriptions, each item of replacement work is measured separately. If a large area of wall tiling is being replaced, for example, the following are measured:

C90.106 Hacking down ceramic wall tiling - m^2
M40.048 Wall: white tiles: exceeding 300 mm wide - m^2
M40.053 ADD where not exceeding 300 mm wide - m^2 (if required)
M40.060 Jointing to existing - m

If an area in the size range 1 to 5 m^2 stated in the "ADD where in repairs" descriptions is being replaced, the following are measured:

M40.048 Wall: white tiles: exceeding 300 mm wide - m^2
M40.055 ADD where in repairs 1 to 5 m^2

The General Directions, page 4, state that rates for "work in repairs" throughout include:
"For removing any existing work.
For new work to match existing.
For all preparatory work and making good.
For jointing new to existing work.
For work of any width."

Q. Do the rates include for excavating by hand or machine?

A. The Preambles to items D20.008 to D20.028 are not intended to imply that excavation by hand is equal to excavation by machine, but is intended to acknowledge that most excavation consists of a mixture of excavation by hand and machine. The rates are built-up from first principles using prices and labour constants for both machines and hand labour. The actual mixture will vary from job to job, but the rates in the Schedule, by its nature, will have to be an average.

In some circumstances, where excavation is required to be carried out by hand, the rates are multiplied by 2 in accordance with D20.029.

The intention of this item is best shown in the following examples:

1. A contractor is instructed to carry out excavation and chooses to excavate by hand because he has spare labour and does not want to hire a machine for a short period. In this case D20.029 does not apply.

2. A contractor is instructed to carry out excavation in a location where access by machine is impossible. An instruction to excavate by hand is implicit and D20.029 does apply.

3. A contractor is instructed to carry out excavation in a location where access by machine is possible but not desirable and the instruction specifically orders hand excavation. An instruction to excavate by hand is explicit and D20.029 does apply.

Q. Under what circumstances is Earthwork Support measured?

A. The method of measurement for Earthwork Support is the full depth to all vertical and sloping earthwork faces exceeding 45° from horizontal, whether or not support is required, except to faces not exceeding 0.25 m high. This applies to all faces and is strictly a *measurement* rule that has no bearing on operational practicalities.

Q. Item G20.006, Structural sawn softwood, wall or partition member is measured "per metre – per 10,000mm^2 of sectional area (which is equivalent to a 100 x 100mm section of timber). How is a 100 x 50mm timber section measured and valued? How is this affected by item G20.012?

100 x 100mm timber will be £9.93 per metre (the sectional area is 10,000mm^2 and G20.012 does not apply as it is exceeding 5000mm^2)

100 x 75mm timber will be £9.93 x (7500/10000) = £7.45 per metre (the sectional area is 7,500mm^2 and G20.012 does not apply as it is exceeding 5000mm^2)

100 x 50mm timber will be (£9.93 + £1.23) x (5000/10000) = £5.58 per metre (the sectional area is 5,000mm^2 and G20.012 *does* apply as it is *not* exceeding 5000mm^2).

Q. When pricing for suspended ceiling works, does the addition for 'work in staircase areas' cover the cost of scaffold, or is this just an addition for the works at height or in a confined area, etc.?

A. All rates include "For providing scaffolding or working platforms not exceeding 4.50 metres above the base of any scaffolding". (see General Directions – Section A).

Q. A project requires the repair of a paved patio area to render it free from trip hazards. The existing paving is in a 'coursed' pattern using three different sizes of flag stone; 300x300, 600x600 & 600x900. In the sketch below, black areas represent 'take up and renew precast concrete paving' and grey areas represent 'take up and re-lay existing precast concrete paving'.

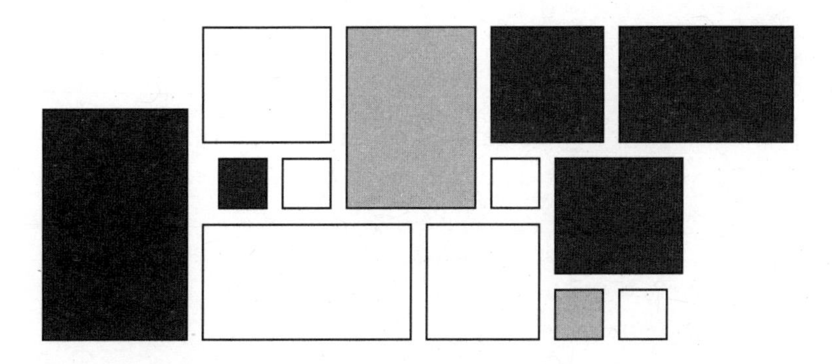

How are items Q25.001, Q25.008, Q25.009 and Q25.010 used for the measurement and valuation?

A. The intention of Items Q25.001, Q25.008, Q25.009 and Q25.010 is as follows:

1. New work exceeding 5 m² is deemed to include all cutting and jointing: measure and value using item Q25.001.

2. New work exceeding 5 m² and abutting existing similar work: measure and value using item Q25.001 and, in addition, Q25.010, where new and existing work meet.

3. Repairs exceeding 5 m²: measure and value using item Q25.001, Q25.010 and, in addition, C90.146, for taking up existing.

4. Repairs not exceeding 5 m²: measure and value using Q25.001 and Q25.008 or 009, as appropriate, but not Q25.010 or C90.146. "In repairs" is deemed to include removing existing work and jointing new to existing (see page 4 of General Directions).

It is intended that Q25.008 and 009 are used for each individual repair, remote from each other. Where, for example, a number of paving slabs not adjacent to each other are required to be taken up and replaced, the areas are not added together to arrive at a total that may exceed 5 m².

In the example, however, renewing and relaying would both require taking up existing and it is suggested that these areas should be added together, where bounded by work not taken up, to produce areas of "repair". Each area of repair would then be treated separately according to 3. or 4. above. Assuming renewing and relaying are carried out at the same time, there is no "jointing to existing" between them.

Q. Is the laying of one large patch of coated macadam valued the same as a number of smaller patches in one area. For example would the laying of a 10m x 5m section be valued the same as 250 patches of 2m² over a large area or would the multiplier factor Q22.002 be applied to the individual patches?

A. See Example BW12.

1. The examples show how to measure the situations described. The first shows an area between 10 and 25 m², the second 500 m² in one area and the third 250 patches at 2 m² each. The difference in cost between example 2 and 3 is the amount of jointing to existing (assuming all new work is meeting existing Macadam around each area). The number of patches executed in one location at the same time does not affect the application of the multipliers.

2. The phrase "executed in one location" means the site of the work identified on the Order or Orders. Note the use of the word "location" as opposed to the use of the word "area" under Method of Measurement for Q22.

3. The phrase "at the same time" refers to work that has been ordered to be carried out and which the Contractor has had sufficient details to plan and carry out in the manner that is most economical to him. If he was issued an Order for some work and then, having arranged labour and materials for that work only, was given some further work in the same location, this could be interpreted as not "at the same time".

Q. Should taking out and re-fixing traps in accordance with C90.168/1 be paid in addition to R11.243?

A. The rate R11.243 is for clearing an obstruction in a trap, waste pipe and overflow pipe. The method of achieving this is at the Contractor's discretion or subject to the nature of the blockage. Sometimes this may be by means of a plunger or sometimes it may be by removing the trap. Removing the trap may, in some cases, actually be easier than prolonged use of a plunger. Whichever method is chosen, the rate is payable for the end result – a cleared trap, waste pipe and overflow.

Taking out and re-fixing traps in accordance with C20.168/1 should not be paid in addition to R11.403. (It should be noted that taking out a WC pan would be payable in addition to Item R11.401, as the method of clearing the obstruction is specifically stated.)

Q. If a drain is rodded in two directions from one manhole or from one manhole through the next manhole and into the next drain run, are rates R12.390 or 391 paid twice?

A. The rates R12.390 and 391 are for "rodding drain from manhole or clearing eye". The process would consist of gaining access, inserting the rods into the manhole, rodding the drain in either, or both, directions and then withdrawing. The rate is payable for each time this process is necessary. If the rods pass through the next manhole into the next drain run (within the stated "Distance between access points"), this is not "rodding drain from manhole" and the rate is not applied a second time. If, for whatever reason, it was not possible to rod straight through the second manhole, then the rods would be withdrawn and inserted into the second manhole and only then would the rate be payable a second time.

The description clearly states that "removing and replacing in position manhole covers or the like" is included.

Q. If dismantling and removing pipework or ductwork, are the Rates D and E applied to all the fittings?

A. Installing pipework or ductwork (Rate B) is measured over all fittings. Installing fittings (Rate B) are measured as "Extra over".

The General Directions, page 1, state that:

"Rates D and E are derived by applying the appropriate multiplying factor to the principal items being dismantled as stated in the particular Section preambles - ancillary and related items not dismantled from such principal items are deemed to be included."

Dismantling and removing for scrap (Rate D) is measured over all fittings, but not the fittings. It is assumed that, in scrapping, the fittings themselves would not be dismantled.

Dismantling and removing for re-use (Rate E) is measured over all fittings. If the fittings are to be re-used in a new location, then dismantling and removing for re-use (Rate E) of the fittings, are measured as "Extra over".

Q. Is connecting (or disconnecting) cables included in the rates for cables or in the rates for luminaries or accessories?

A. Installing luminaires and accessories (Rate B) includes for connecting all cable ends and tails (see "Rates for the following include" on pages 269 and 320 of the Electrical Schedule). Therefore, Rates D and E will include for disconnecting. Only connecting cable ends and tails into existing equipment is measured separately (see page 197).

The Method of Measurement on page 71 states that cleats and hangers, cable terminations and cable jointing are measured separately but that connecting cable ends and tails are included elsewhere within other items (e.g. luminaries in Section Y73 and accessories in Section Y74). Connecting cable ends and tails are not identified in the list of items included because they are included in items in other Sections. It is the items to which cable ends and tails are fixed that include for connection in their Rate B, not the cable itself.

In short, if it is the luminaire, accessory or the like that is being dismantled and removed, then disconnecting is deemed to be included in the Rates D and E. If it is the cable itself that is being removed and the luminaire, accessory, etc. that remains, then items on page 197 (connections to existing equipment), with the multiplying factor applied, would be used.

Q. Does the measurement of electrical work include for testing?

A. Whether new or alterations to existing, work measured and valued in accordance with the rates in the schedule is deemed to include for testing and commissioning. (See General Directions "Rates throughout include", "Rate B", second to last paragraph). Rate B is for "installation", and installation must include setting to work, making safe and complying with the Regulations. It therefore includes for testing. If any part of the existing installation requires testing as a result of the new installation, this is also deemed to be included in Rate B as it is a result of "installation".

Section Y81 is for testing, specifically ordered, of existing installations not affected by any installation work carried out by the Contractor and where there would, obviously, be no Rate B to apply. The only exception to this is the rare case where the Client specifically requests additional testing to new installations, in additional to the basic testing required for setting to work and safety. (See "Notes", first page of Section Y81).